AF540303

Organic Farming for Sustainable Agriculture and Horticulture Production

NIPA® GENX ELECTRONIC RESOURCES & SOLUTIONS P. LTD.
New Delhi-110 034

About the Authors

Prof. (Dr) Rameshwar Dass Gupta (R.D. Gupta) Ex-Associate Dean Cum Chief Scientist Krishi Vigyan Kendra (KVK) of Sher-e-Kashmir University of Agricultural Sciences and Technology (SKUAST) Jammu was born on 27th September, 1941 at village Thub, Tehsil, Jammu, J&K State, now Union Territory. He earned his B.Sc. Agriculture Degree during 1965 with specialization in Agricultural Chemistry and Soil Science from Jammu and Kashmir University securing 3rd position. After serving about six months from October, 1965 to March, 1966 as a teacher teaching Chemistry to the final year students of Higher Secondary in the Ranbir Higher Secondary School Parade Ground, Jammu, he got admission in M.Sc. Agriculture (Agricultural Chemistry and Soil Science) at Agriculture College Kanke, Ranchi, Bihar, now Jharkhand during April, 1966. He completed the aforesaid degree after getting through a number of courses relating to Soil Physics, Soil Chemistry, Soil Genesis and Classification, Soil Microbiology as well as Plant Bio Chemistry, Plant Physiology and Physical Chemistry. He secured 2nd position after completing aforesaid courses during June, 1968.

Dr. Gupta has resorted his M.Sc. "Agriculture Thesis" work under the guidance of Dr. K.K. Jha, the then Soil Chemist Bihar, Agricultural Research Institute, Kanke, Ranchi, who later on became Vice Chancellor of Rajindra Agricultural University, Bihar, India. His thesis having title, "Genesis, Physical, Chemical, Mineralogical and Microbiological Nature of the Soils of Jammu and Kashmir". Thus, he was one of the pioneers who worked on the soils of Jammu and Kashmir, now, Union Territory. He did Ph.D. in Agriculture with specialization in Soil Science and Water management in 1980 from Chaudhary Sarwan Kumar, Himachal Pradesh Krishi Vishva Vidyalaya (CSK HPKVV) Palampur, India. It is further added that Dr. R.D. Gupta has studied special courses such as "General Bacteriology, Inorganic Nutrition, Chemistry of Soil Organic Matter and Analytical / Physical Chemistry during his Ph.D. in Agriculture". His Ph.D. thesis work relating to "Mineralogical and Microbiological properties of soils Kangra District (Himachal Pradesh) developed under different Bio, Climo and Litho sequences", was done under the able guidance of Dr. B.R. Tripathi Dean, College of Agriculture, under HPKVV, Palampur who later on became Vice Chancellor of Acharya Narendra Deva University of Agriculture and Technology, Kumarganj, Ayodhya, Faizabad, U.P. Apart from the above, Dr. R.D. Gupta also holds Post Graduate Diploma in Ecology and Environment with a specialization on the "Wildlife of Pirpanjal Himalayan Region of Jammu and Kashmir", from Indian Institute of Ecology and Environment, New Delhi.

Author's first book relating to "Problems and Management of Soil and Forest Resources of North West Himalayas" which covers a number of topics besides "Genesis and Classification of Soils, their Physico-chemical and Mineralogical

Properties" was published during the year, 1991. Second and third Book entitled "Environmental Degradation of Jammu and Kashmir Himalayas and their Control", and "Environmental Pollution - Hazards and Control" were published during 2005 and 2006, respectively, which relate to Environment and Ecology, besides pertaining to various Soil Science Topics. Thus, the author is not only a leading "soil scientist having specialization in soil microbiology and pedology but is also a famous "Environmentalist and Ecologist". Other published books on the name of Prof. (Dr) R.D. Gupta are "Glimpses of North West Himalayas, 2009", "Wildlife and Wetland Ecosystems of Jammu and Kashmir Himalayas, 2016" and "Agro-techniques and Uses of Medicinal Plants 2016". The above quoted book relating to "Medicinal plants" is co-authored one, written by Dr. R.D. Gupta, S.K. Gupta and S.D. Bhardwaj. Another co-author's published Book is "Integrated Nutrient Management (INM) in a sustainable Rice-Wheat Cropping System", authored by Anil Mahajan and R.D. Gupta which was published by Springer Science-Business Media B.V. 2009.

He started his service career as an "Assistant Extension Specialist, Soil Science", after joining Punjab Agriculture University (PAU) at Dharmsala, District Kangra, HP during December 1968. Later on, he was adjusted in CSK, HPKVV, Ralampur and served in various capacities such as Assistant Scientist / Assistant Professor, Associate Professor/Extension Specialist cum Incharge of First Krishi Vigyan Kendra (KVK) at Daula Kuan, District Sirmour, Himachal Pradesh during May, 1983 and worked there up to 21st February, 1985. On 22nd February, 1985, he joined as Deputy Director Extension Education at Regional Agricultural Research Station (RARS) R.S. Pura, under SKUAST (Sher-e-Kashmir University of Agriculture Sciences and Technology). During 1992, he became Chief Scientist and Head of SKUAST, R.A.R.S, R.S. Pura as well as Chief Scientist of first KVK opened at R.S. Pura, Jammu District of Jammu region of Jammu and Kashmir State, now the Union Territory. After creation of the separate Agricultural University for Jammu Province, named as Sher-e- Kashmir University of Agricultural Sciences and Technology, Jammu (SKUAST-J), Dr. R.D. Gupta was appointed as the "Founding Associate Dean", Faculty of Agriculture SKUAST-J. Besides teaching work both of graduate and post graduate students in agriculture, he was also assigned to look after the work of all "Research Stations" viz. Pulses Research Station Samba, Dryland Agriculture Research Station Rakh Dhiansar, Horticulture Research Station Udhewala, Soil and Water Management Research Station Ponichack as well as Agricultural Research Stations located at Rajouri and Poonch. He was also the Head, Division of Agricultural Chemistry and Soil Science and Agroforestry.

Besides teaching Soil Science and Agricultural Chemistry and Environment Ecology to graduate and Post graduate students, he has guided a number of students of M.Sc. Agriculture (Agricultural Chemistry and Soil Science) as well as those of

Ph.D. He has organized a large number of professional training programmes not only in Agriculture but also in Horticulture, Floriculture, Animal Husbandry as well as in Home Science under Krishi Vigyan Kendras (KVKS). Dr. R.D. Gupta is recipient of US Dollars 50 for his best paper among the participants for USDA and SAARC Journals. He has been awarded by the Central Mahajan Sabha Jammu and Jammu and Kashmir Women Welfare Society for significant contribution in Agriculture Research during 1997-1998 and transfer of Agricultural Technology during 2001, respectively. He was chosen and conferred with an honorary appointment to the Research Board of Advisor by an American Biological Institute (ABI) USA in 1999. An award was provided by Soil & Conservation Society of India in 2006 for outstanding contribution in the area of Teaching, Research and Extension. He was further awarded by Bharat Vikas Parishad during 2009 in the field of Agricultural Chemistry and Soil Science. He was also selected and bestowed upon with "Gold Medal Plaque" by the ABI during 2010 for the best work done in Agricultural Extension, Research and Teaching in Agricultural Chemistry and Soil Science.

"Life Time Achievement Award 2019" was given with great honour to Dr. R.D. Gupta for his highly significant contribution in the field of Natural Resource Conservation and Management in the Hilly Region of North West India by Soil Conservation Society of India (SCSI).

Best Citizen of India Gold Medal Award" was presented to Prof (Dr.) R.D. Gupta, Excellence in his respective field on the occasion of 73rd National Unity Conference for "Individual Achievement and National Development" on 22nd February 2020, at Chennai", by Global Economic Progress and Research Association (GEPRA, New Delhi).

"Lead Speaker Award" was provided by the Head of Division of Agronomy, Faculty of Agriculture SKUAST-J. It was after delivering a lecture on, "Organic Farming" to the Trainees in the ICAR sponsored short course entitled "Conservation Agriculture Practices for Enhancing Productivity on Resource Use Efficiency in Major Cropping systems organized from 4th to 13th February 2020".

"Mahatma Gandhi ji Gold Medal Award" was presented to Prof. (Dr.) R. D. Gupta, excellence in his respective field on the occasion of 75th National Unity Conference on Individual Achievements and National Development on 30th October, 2021 Bangalore by GEPRA, New Delhi.

Dr. Gupta is a life member of a number of societies like Indian Society of Soil Science, Soil and Water Conservation, Clay Mineral Society of India, the Indian Society of Soil Survey and Land Use planning, Indian Society of Tree Sciences, Indian Society of Environment and Ecology, Society of Environment and People, Association of Rice Research Workers, Indian Farmers Digest, Agrobios

Newsletter, Down to Earth and Ayurveda for Holistic Health, Association of Rice Research Workers.

He has more than 150 peered review publications. More than 100 research papers are in various Indian Journals of repute as stated above. Twenty-six book chapters are in different books edited by Distinguished writers like Late Dr. S.K. Chadha Ex- Professor Geology, Dr. B.L. Kaul Ex-Head, Department of Zoology, Govt. College for Women, Parade, Jammu.

Apart from the above, book chapters of Dr. R.D. Gupta also stand appeared in "Dryland Farming in India: Constraints and Challenges". "Bio-Industrial watershed Development", edited by Dr. J.L. Raina, Head of the Department of Geography,

G.G.M. Science College Jammu and Suraj Bhan (President) and his associates (S. Subramaniyan, V.K. Bharti, R.L. Karale and Shamsher Singh-Members) of Soil Conservation Society of India, New Delhi. A few book chapters relating to plant nutrients both macro i.e., primary (N, P,K, Ca, Mg, K) and Micro (Fe, Mn, Cu, Zn, B and Mo) in various oil seed crops for mixed cropping and / or intercropping systems have been detailed in the books written by Dr. S.K. Gupta, Ex-Director Education of SKUAST-Jammu. These books have been published by Springer Science + Business Media, LLC, 2012 and LLC, 2016.

Served as referee of Himachal Journal of Research, Journal of Clay Research and Journal of Soil and Water Conservation. Remained Paper Setter for Soil Agricultural Research Service (ARS) for a number of years. Similarly, he remained paper setter for Kashmir Administrative Service (KAS) in Agriculture. He is the Vice President of J&K Paryavaran Sanstha (Environment Society) and is also the Member of J&K, Pollution Control Board.

Dr. S.K. Gupta, born in 1959, Former Professor and Head, Division of Plant Breeding and Genetics/Director of Education, SK University of Agricultural Sciences and Technology, FOA, Chatha, Jammu, India, and holds a brilliant academic and service record. For almost three decades he has devoted his research interests to the area of oilseed brassicas.

Dr. Gupta obtained his post-graduate degrees (M.Sc., Ph.D.) from Punjab Agricultural University, Ludhiana, India, in 1984 and 1987, respectively. He is the recipient of a post-doctoral fellowship in plant biotechnology, and has published more than 100 research papers in esteemed national and international journals, mostly on Brassicas. He has already developed eight varieties of rapeseed-mustard. In addition, he has written two books on plant breeding and edited eight books on Rapeseed Breeding-Advances in Botanical Research, Vol. 45, Academic Press, Elsevier Publishers. For his excellent scientific endeavours, he has been conferred with the Young Scientist Award: 1993-1994 by the State Department of Science and Technology.

Organic Farming for Sustainable Agriculture and Horticulture Production

R.D. Gupta
Ex-Associate Dean Cum Chief Scientist
Krishi Vigyan Kendra (KVK)
Sher-e Kashmir University of Agricultural Sciences and Technology
Jammu, Jammu and Kashmir, India

S.K. Gupta
Former Professor and Head/Chief Scientist (Oilseeds)
Division of Plant Breeding and Genetics,
Sher-e-Kashmir University of Agricultural Sciences and Technology
FOA, Chatha, Jammu
Jammu and Kashmir, India

NIPA® GENX ELECTRONIC RESOURCES & SOLUTIONS P. LTD.
New Delhi-110 034

NIPA® GENX ELECTRONIC RESOURCES & SOLUTIONS P. LTD.

101,103, Vikas Surya Plaza, CU Block
L.S.C.Market, Pitam Pura, New Delhi-110 034
Ph : +91 11 4386 0225, 9717133558, 9540816132
E-mail: newindiapublishingagency@gmail.com
Website: www. niparesources.com

Print ISBN: 978-81-19002-54-2

ebook ISBN: 978-81-19002-54-0

Composed and Designed by NIPA®.

Preface

The growing awareness about the bad and ill effects of the synthetic chemicals-pesticides, insecticides, fungicides, bactericides, herbicides etc. and fertilizers-nitrogenous, phosphatic and potassic ones in the chain which stand mentioned below in details paved the way for "Organic Agriculture" or Organic Farming. It is a holistic production management system that promotes and enhances health of agroecosystem, including biodiversity, biological cycle and biological activity. It is point to mention that in the developed countries, the fertilizer was used judiciously- application of the Nitrogenous(N), Phosphatic (P_2O_5) and potassic (K_2O) fertilizers was made in right proportion by maintaining their proper proportion i.e.,4:2:1 ratio. However, in Indian conditions, this ratio (National Average) has been worked out to be 7:6:3:7:1 taking into consideration of number of years. Thus, indiscriminate use of nitrogenous fertilizer without phosphate and potash has damaged the productivity of agricultural land in the long run and also created a number of health hazards both to the soils, human beings and animals. In the major rice-wheat growing regions of Punjab and Haryana the organic carbon content declined which exhausted the secondary (Ca, Mg, S) and micronutrients, (Fe, Mn, Cu, Zn B, Mo). Their deficiency affected the yield of the crops and caused a number of disorders in plants and animals that feed on them and even the human beings. There was also a loss of micro-organisms in the soils which as a result inhibited their ability to absorb conventional fertilizers.

Apart from this, use of nitrogenous fertilizers alone produced negative effects and proved more disastrous in acidic soils. It also became potent source of ground water contamination in the form of nitrate. Drinking of nitrate contaminated water caused 'Methemoglobinemia 'disease. Some nitrosamines formed by nitrate reaction with secondary amines were suspected too carcinogenic. Other hazards associated with indiscriminate use of nitrogenous fertilizer were cause of health problems like degeneration of vascular diseases of lungs, liver and heart when fodders like oats, maize stalks raised on excessively nitrogenous fertilized soils.

Heavy metals like Cd, Cr, Ni and Pb accumulated in soils and ground water due to excessive use of fertilizers have caused cancer esophagus, breast and uterus.

The presence of fluorine above permissible limit (1-5pm) has been found in many soils. The main source of fluoride pollution was phosphatic fertilizer

Apart from this, eutrophication of waterbodies formation of carcinogenic agents and depletion of ozone layer were the other effects produced by the injudicious use of chemical fertilizers. In light of the above, the book entitled Organic Farming for Sustainable Agriculture and Horticulture Production written by Prof (Dr) R.D. Gupta and Dr S.K. Gupta, will prove a boon not only to farming community but also to the Agricultural production concerned. The book comprises of thirteen chapters. Chapter one deals with need of organic farming and chapter two and three with history and importance of organic farming or biodynamic farming in India. Introduction of organic farming and horticulture and its relevance in India particularly, Jammu and Kashmir has been provided in the chapter fourth of the book. Chapter five has been dealt with production of vegetables using through various organic sources. Organic farming and cultivation of medicinal plants, have been stated in chapter six. In chapter seven, the ways of achieving food security have been mentioned. How rural women assist in sustainability of agriculture through organic farming, has been mentioned in chapter eight. In the chapter nine and ten of the book, how rose flower and pulses can be grown organically have been given. Organic farming and land degradtion has been provided in chapter eleven; the chapter twelve speaks about protecting the degraded land of the Indian Himalayan region through organic farming. Chapter thirteen relates to how, agroforestry connects with organic farming, has been detailed in respect of various Agroforestry systems i.e. Agriculture- silviculture, Silviculture — Agriculture- Animal Husbandry, Silviculture — Horticulture, Silviculture — Agriculture — Horticulture — Fishery etc.

Authors

Contents

Glossary

Absorption: It is the mode of absorption of plant nutrients through an ion exchange process

Acidification: It is one of the soils forming process by which acid soils are formed

Acidity: Acidity is either characterized due to the presence of free acids or due to the lack of basic material. The former is called positive acidity and the latter is known as negative acidity

Acid soils: These are the soils which show low pH in the range of 4.0 to less than 7.0

Actinomycetes: Actinomycetes are unicellular like bacteria. They are also called thread bacteria owing to the presence of thread like structures or mycelia in their body.

Adsorption: It is the surface phenomenon by which a substance accumulates on the surface of another

Aerobic and Anaerobic Bacteria: All soil bacteria require for their growth a certain amount of O2. However, some of them, can continue their activity with much less content of O2 (oxygen] than others. Those bacteria which require an abundant of O2 supply than others are known as "Aerobic bacteria". Contrary to this, those which prefer less oxygen supply for their growth, are called "anaerobic bacteria".

Algae: Algae are a group of unicellular or multi cellular or simple plants containing chlorophyll. They are represented by green algae, blue green algae and diatoms.

Alkali and Saline Soils: They contain high concentration of soluble salts but exchangeable Na+ is more than 15me in alkali soils.

Alluvial soils: These soils are formed by the deposition of sand, silt, and clay brought about by the rivers including their tributaries. Out of all the soil groups of India, alluvial soils are the most extensive and very important.

Aminization: As a result of enzymic hydrolysis affected by the microorganisms, the complex nitrogenous compounds (Proteins) decompose into amino acids which is designated by the word aminization.

Ammonification: Amino acids or amino compounds when are converted into ammonium form, is known as ammonification. Ammonization proceeds the best in well drained aerated soil. The fate of ammonical nitrogen is three-fold. Firstly, some of it is utilized by the ammonifying and other organisms themselves, which are capable of using this kind of nitrogen. Mycorrhizal fungi are also able to use ammonical nitrogen and transfer it to the host plants. Secondly some higher plants are able to use this type of nitrogen. Thirdly, it is oxidised by certain special kind of bacteria known as nitrifying bacteria (Nitrosomonas, Nitroso coccus) to nitrate, this process is generally known as nitrification.

Amoeba: Amoeba is the protozoa which is unicellular microorganism belonging to animal kingdom. It is one of the members of the class "Sarcodina" and moves by means of temporary protoplasmic extrusions from the cell body.

Amphiboles: A complex group of primary minerals consisting of magnesium calcium iron silicates which are to not very resistant to decomposition.

Animals in Soils: Animals in soils are characterized by the presence of both macro and micro. The macro-animal or micro-organisms of the soils are chiefly rodents, millipedes, centipedes, insects, worms etc. Of the abundant microscopic animal life in soils, two groups are particularly very important. These are: i) Nematodes and ii) Protozoa. Protozoa are the most varied and numerous of the micro-animal population. As many as 250 species of protozoa have been isolated. They are the simplest form of animal life. They are one celled organism.

Anion exchange in soils: The chemistry of anionic exchange is not fully known. It is, however, certain that anionic exchange does take place in soils, although to a much lesser extent than cation exchange.

Autotrophic and Heterotrophic Bacteria: Those bacteria which can live partly or wholly without organic matter content are called as "Autotrophic Bacteria". Although these bacteria are very small in number yet a remarkable group of these bacteria obtain their energy by the oxidation of inorganic substances such as ammonium (NH4], nitrite (NO2), sulphur and ferrous compounds etc, and derive their carbon from carbon dioxide. On the other hand, those bacteria that require their energy and carbon from organic sources are called, "Heterotrophic Bacteria". It is remarkable to note that most of the soil bacteria belong to this class. As the Bacteria are lacking of chlorophyll and multiply mostly by fission, and are, therefore, known as Procaryotes.

Autotrophic organisms: These are the organisms which can live partly or wholly without organic matter.

Azofication: The process of non-symbiotic nitrogen fixation in soil is known as Azofication.

Azotobacter: Aerobic bacteria of the genus, Azotobacter of which five species have been recognised. These are as follows: i) Azotobacter agilis ii) Azotobacter beijerinckia iii) Azotobacter chroococcum iv) Azotobacter nigricans v) Azotobacter vinelandi

Bacteria: The term bacteria include a large group of typically unicellular microscopic organisms widely distributed in air, water, soil, the bodies of living plants and animals and dead organic matters. The bacteria are placed in that division of the plant kingdom which is known as the Protophyta. This division contains 3 classes and 12 orders.

Biofertilizer: A carrier-based preparation having different kinds of bacteria like Azotobacters known as Azotobactrin or Azotobacter culture and Rhizobium culture, Phosphate Dissolving Bacterial culture Azosopirillum, Blue Green Algae i.e., BGA, as well as Azolla (An aquatic, floating fresh water fern capable of fixing atmospheric nitrogen freely) are other biofertilizers.

Biogas slurry: The residual slurry that comes out of the digestion tank biogas plant is called the biogas slurry. On an average, it possesses 1.8% N, 41.1% P2 O and 1.5% K 0.5 2

Blood meal: It is a product of slaughter house which is found to have 10-12% available N, 1.0 to 1.5% P and 1.0% K. Blood meal has been proved a very quick acting organic manure.

Blue GreenAlgae (BGA): The blue green algae are also called cyanobacteria and are unicellular like bacteria belong to a heterogenous group of prokaryotic photosynthetic organisms which contain chlorophyll 'a'. They are nonsymbiotic, free living atmospheric nitrogen fixer that grow very well in water logged soils like rice fields. Apart from fixing atmospheric nitrogen, BGA produce growth promoting substances which help to have better growth of rice plants.

Bone meal: Bone meals should be applied at the rate of 5 to 15 tons per hectare. They are more textured, loose soils than on compact clayey soils. It is point to mention that bone meal is used as a manure from time immemorial, chiefly as a source of phosphorous and partly also for nitrogen.

Bulk density: Bulk density is the weight or mass of a unit volume of a dry soil and is determined by dividing the dry weight of a given volume of a soil as it occurs in the natural structural condition by the weight of an equal volume of water. Or the ratio of the mass of the soil solid (over "dried soil) to the bulk volume of the soil and determined by Core Sampler method.

Bulky organic manures: Manures prepared artificially from plant residues and animal waste products are called bulky organic manures. They contain very small amount of plant nutrients and large quantity of organic matter.

Calcareous Soil: A soil containing sufficient amount of free calcium carbonate and gives effervescence when treated with cold 0.1 N hydrochloric acid.

Calciphytes: Plants which require lot of calcium are known as calciphytes.

Capillary (Conductivity): (i) (Qualitative). The physical property relating to the readiness with which unsaturated soils transmit water, (2) (Quantitative). The ratio of the water flow velocity to the driving force to unsaturated soil.

Carbohydrates: They are a large group of compounds containing C, H and O only. They are produced by the photosynthesis process which occur in plants and are essential in metabolism in all living organisms.

Carrier: A substance which possesses high organic matter accompanied with having higher water holding capacity and always supports the growth of organism is known as carrier. Lignite, peat, charcoal etc. are suitable carrier materials for biofertilizers.

Cation Exchange Capacity (CEC): The sum total of exchangeable cation that a soil can adsorb which is expressed in centimeters per kilogram of soil is called CEC.

Chelates: The word chelates stands derived from Greek word Chela which means a crab's law. They contain organic metallic molecules of varying sizes and shapes in which the functional groups of organic matter bind the nutrients in a ring like structure by means of multiple chemical bonds.

Chlorophyll: A green colouring matter or pigment found in many of the green plants which is essential for the process of photosynthesis is called chlorophyll.

Climate: Climate is an aggregate of weather conditions over a long period of time. Climate includes the following five elements:

1) Temperature
2) Rainfall
3) Humidity
4) Pressure
5) Wind

Clod: A natural lump of soil that exists as an isolated entity in the field or formed artificially by soil disturbance like ploughing digging etc., is a clod.

Clostridium (Plural Clostridia): These are anaerobic bacteria which can tolerate more acidity than Azotobacter group of bacteria and as such are wider spread.

Compost: Compost is an organic manure similar to farm yard manure in properties. It is artificially made by the decomposition of plant residues under the action of micro-organisms. The raw material for compost is the farm straw, weeds, sweepings, leaves, stubbles, rotten vegetables and fruits etc., that contain about 60 per cent cellulose, 15 to 20 per cent lignin, 15 to 20 percent water soluble substances and 1.2 to 3.0 per cent proteins.

Composting: it is a biological process by which both kinds of micro-organisms (aerobic and anaerobic) decompose the organic matter to lower down the C:N ratio of the refuse.

Decomposition: Process of the chemical break down of minerals and organic compounds into simpler components accomplished with the help of microorganisms is known as decomposition.

Denitrification: This is the reverse of nitrification where in the nitrates of the soil is reduced to nitrites, oxides of nitrogen, ammonia and nitrogen gas.

Dispersion: The process of movement of biofertilizers/fertilizers applied to the soil for short and long distances is called dispersion.

Dominancy of Nematodes: Out of the three groups of nematodes as stated above, the first group is by far the most numerous in the soil. The second group is the most important agriculturally.

Drainage: The removal of excess surface water or ground water from land is known as drainage.

Earthworms: The most important of macro-animals of the soil are earthworms, of which there are a number of species. The amount of soil that these creatures pass through their bodies annually may amount to as much as 37.5 tonnes earth per hectare.

Electron microscope: It is used for seeing the viruses which are the smallest organisms even smaller than bacteria which cannot be seen with ordinary microscope.

Endomycorrhizal: An association between fungi and roots of a plant in which fungal hyphae infect the roots and remain up to the critical region (parenchyma of roots) by secreting cellulosic enzymes. They are useful particularly for citrus plants and deciduous shrubs.

Environment: The word environment refers to surroundings which varies from place to place and physiography, topography, climate and natural resources (land/soil, air, water, forests etc.).

Enzyme: A component of protein which catalyses or speeds up a specific chemical reaction, is called enzyme.

Escherichia coli: A bacterium of exclusively faecal origin which forms 90 per cent of coliform group of bacteria in the human intestines is called Escherichia coli.

Eutrophication: In eutrophication process wetland becomes rich in plant nutrients like N,P,K and others which support a dense plant growth population especially of algae where under the other population of plants and animals is badly affected.

Evergreen: A plant or a tree which always bears leaves throughout the year is called evergreen. As for example, pine tree.

Farm Yard, Manure (FYM): It is a bulky or general organic manure supplying all the tree constituents of plant food. i.e. N, P and K as well as other nutrients although in low amount. FYM is a mixture of solid and liquid excreta of farm animals (Dung and urine) along with litter (bedding material) and left over materials from roughages or fodder of cattle.

Fauna: It refers to animals of a region consisting of both kinds viz. higher and lower animals' kingdom.

Fertilizers: These are artificial or chemical manures added to the soils or applied directly to crop foliage to supply plant nutrients for the growth of plants.

Fungi: Any tiny aerobic, heterotrophic protists containing no chlorophyll, many celled, is called fungus, plural fungi.

Fungicide: Any pesticide which inhibits or prevents fungus growth causing disease in plants is called fungicide.

Garbage: Animal, plant, fruit, or vegetable residues resulting from the handing, preparing and cooking of food, is known by the term garbage.

Grassland: Land which is dominated by herbaceous vegetation and grasses of various species.

Green manure: Any green crop that is buried directly into the soil to increase its fertility and to improve its properties.

Growth: An irreversible process in which there is an increase in size, shape, and volume of an organism is known as growth.

Habitat: Any physical part of an environment which is inhabited by an organism or population of organisms, is known as habitat.

Haemoglobin: A respiratory pigment that is found in the blood plasma is known as haemoglobin.

Herbicide: Any chemical compound that is used to destroy off plants or weeds is known by the word herbicide.

Humification: It is the process by which the organic matter is changed or transformed into humus.

Humus: It is well decomposed organic matter. The organic matter in soil is derived mostly from plants and microorganisms, only a small fraction comes from animals. Humus is of a colloidal nature like clay and is dark brown or black brown in colour.

Hydraulic conductivity: The ability of soil to conduct water and is expressed as length per unit time or an expression of the readiness with which a liquid such as water flows through soil in response to given potential gradient.

i) Those living on decaying organic matter, ii) Those parasites on higher plants and roots, and iii) Those predatory on other nematodes and small earthworms. Many nematodes can resist drought by passing into a resting stage, when they become coiled into simple shapes.

Incineration: A method of treating a refuse to reduce its volume and weight to have an innocuous residue is called incineration.

Indigenous: Native or original to an area, not introduced from outside is represented by the word indigenous.

Inhibitor: It is the opposite of catalyst as it retards the rate of chemical reaction.

Inoculation: Treatment of soil or seeds with bacterial or organism culture is known as inoculation.

Inoculum: When bacteria or other organisms are added in compost making or the soil to begin with biological action is called inoculum.

Inorganic Fertilizers: Fertilizers having a mixture of chemical compounds which are added into the soil in order to improve its fertility is known by the word inorganic fertilizers. However, the use of chemical fertilizers in large amount have now proved harmful to environment.

Integrated Nutrient Management System (INM): It is a system which aims at improving and maintaining soil fertility and productivity by using suitable organic and inorganic plant nutrients required for crops growth and quality of farming in an integrated manner.

Intensive Farming: A farming system in which maximum number of crops are grown per year on the same piece of land with an objective to get maximum income per unit area is called intensive farming.

Intercropping and Mixed Cropping: Mixed cropping refers to the system of growing two or more crops in the same field or area with no row arrangement, whereas intercropping represents growing of crops in the same area in separate rows.

Irrigation: The artificial supply of water to the soil for the purpose of providing water to the growing plants/trees, is called irrigation.

Kinds of Nematodes: They fall roughly into three groups on the basis of their food requirements:

Land: That part of the surface of lithosphere which is not usually covered with water and is liable to form the total natural and cultural environment suitable for production of crops, fruit trees and vegetables etc., is known by the word land.

Leaching: It is the process of removal of plant nutrients from the soil by the passage of water through the soil. This is one of the ways by which plant nutrients are lost from the soil.

Leghaemoglobin: It is red pigment like haemoglobin which is found in the human's blood. Leghaemoglobin occurs in the root nodules of leguminous plants. Due to the presence of leghemoglobin in nodules their colour becomes pink and as such they are known as affective nodules. Contrary to this, the nodules which do not contain leghaemoglobin, they are called ineffective nodules.

Legume: It is a leguminous plant. Leguminous plants are very important from agricultural point of view. It is because leguminous plants fix atmospheric nitrogen through rhizobia confined to the root nodules of these plants.

Lichen: It is an association of blue green algae or green algae with the fungi.

Lodging: Falling of crop plants under the influence of heavy rains or wind blowing is called lodging.

Lowland rice: Rice grown on flooded flat land with controlled irrigation called lowland rice. This is also known as irrigated rice or water-logged rice. Lowland rice is generally flooded when the seedlings which are 25-30 days old.

Microbe: Any tiny plant or animal which can be seen with the help of microscope, is called microbe.

Microfauna: Microscopic animal like protozoa and nematode come under the group of micro fauna.

Microflora: Microscopic plants like bacteria, actinomycetes and yeasts as well as single celled algae i.e., blue green algae belong to microflora.

Microorganisms: These are either microscopic plants i.e., bacteria, actinomycetes, fungi and or animals-protozoa and nematodes.

Mycorrhizae: A symbiotic association of fungi with roots of vesicular plants, is called mycorrhizae. It is generally two types - ectomycorrhiza and endomycorrhiza.

Myxobacteria: These are such kind of bacteria which are characterized by the production of slime. They are commonly classified into bacteria, but they have a more complex organization and are probably like the actinomycetes (Thread bacteria). They lie between bacteria and fungi. Some of them readily decompose cellulose aerobically, obtaining their nitrogen from nitrate and ammonium ions.

Nematodes: Nematodes are also known as eelworms. They are found in almost all soils. They are round or spindle shaped in form, one end usually being acutely pointed. In size they are microscopic, seldom being large enough to be seen readily with the naked eye-0.5 to 1.5 mm long.

Nitrification: This is the name given to the process comprising of oxidation of ammonical nitrogen to nitrate nitrogen. The above said process takes place in two coordinated steps, each step being worked out by a separate group of bacteria. In the first step ammonical nitrogen is oxidized to nitrite by Nitrosomonas and then from nitrite to nitrate by Nitrobacter.

Organic Farming: Organic farming may be defined as the system of farming which totally avoids the use of chemicals i.e., chemical fertilizers, pesticides as well as growth regulators and livestock feed additives.

Organic Fertilizer: Any fertilizer which stands derived from animal products like bone meal, dried blood etc; and or from plant residue such as cotton seed cake, neem seed cake, groundnut seed cake is called organic fertilizer.

Organic manure: These are the manures which are mostly made up of dead plant residues and animal remains. They are added to the soil specially for the nutrition of plants and are generally of two kinds viz. Farm Yard Manure (FYM) and Compost.

Organic Matter: The soil organic matter consists of several products ranging from plant and animal residues at various stages of decomposition and excretion of all the organisms including microorganisms living in the soil.

Phosphate Solubilizing Microorganisms: All microorganisms which bring about solubilization of insoluble phosphate (tricalcium phosphate, rock phosphate etc.) into soluble form by formic acid, acetic acid, lactic acid, propionic acid etc.

Phytohormones: Plant growth substances which are produced by the plants are called phytohormones.

Plant Growth Regulators: Organic compounds occurring naturally in plants as well as synthetic materials other than nutrients, which in small amounts promote or modify physiological process in the plants are known as plants growth regulators.

Population of Nematodes: More population of nematodes occurs in soils having more organic matter. They are more numerous in summer than in winter. The phylum belonging to the protozoa possesses the primitive unicellular organisms ranging in size from several microns up to one or more centimeters.

Press Mud: A by-product of sugar industry which is used as organic manure is called press mud.

Primary Nutrients: Nutrients which are required by plants in large quantity such as C, H, O, N, P, K, Ca, Mg and S, are known as primary nutrients.

Secondary Nutrients: Ca, Mg and S are next in importance to other major nutrients such as C, H,O, N, P, and K which are known as primary nutrients, whereas Ca, Mg, and S are known as secondary nutrients. However, like C, H, O, N, P, and K secondary nutrients are also included in the major group of plant nutrients.

Slow-release Fertilizers: Fertilizers whose nutrients are present as a chemical compound or in a physical state such that the availability to plants is slow but uniform over a period of time such as Sulphur coated urea, neem coated urea, urea super granule etc.

Sludge: The solid portion of sewage system of sanitation is called sludge. On an average it contains 1.5 to 3.5% N, 0.75-4.0% P2 O5 and 0.3-0.6% K2O Soil: Soil is the upper most crust of the earth which is capable to grow plants life. As is evident, the word soil is composed of 4 letters viz; s, o, i and 1 which indicate, soul of infinite and life, respectively. Thus, soil is considered a soul of infinite life.

Soil Alkalinity: The degree of intensity, of alkalinity of a soil, can be expressed by a pH value greater then 7.0, is due to an excess of hydroxyl ions or having high exchangeable sodium content or both.

Soil Erosion: The process of detachment of soil, and its transportation by natural agencies or processes such as rain water or wind is called soil erosion. Soil fertility: The inherent capacity of soil to produce crops by way of supplying an adequate amount of available nutrients in balanced form to plants necessary for their growth as soil fertility.

Soil Fexture: Texture of soil represents the physical makeup of the soil and refers to the proportion of different sized soil particles groups or separates in the soil on percentage basis.

Soil Health: The capacity of soil to function within the ecosystem boundaries to sustain biological productivity, sustain environment quality and promote plant and animal health is termed is soil health.

Soil Management: It is the manner in which soil is utilized to produce food, fodder, fiber, and forage crops.

Soil Moisture Regimes: The presence or absence of water in a soil at different times of the year is the soil moisture regime. It is a partial function of climate, soil and landform.

Soil Pollution: Soil pollution is also called land pollution which is the direct effect of chemicals- fertilizers, pesticides and unscientific use of organic wastes and improper drainage.

Soil Porosity: The volume percentage of the total soil bulk which is not occupied by the solid particles or percentage of pore space is called soil porosity.

Soil Productivity: The capacity of the soils to produce crop yield per unit area under a definite set or a specified system of management practices is known as soil productivity.

Soil Profile: The vertical section of soil is called soil profile which is represented by various soil layers or soil horizons depending up on the soil formation.

Soil Structure: The arrangement of soil particles and their aggregate formation into well-defined pattern is known as soil structure. It is characterized by four main forms viz. prism like, plate like, block like, and spheroidal structure.

Soil tilth: Soil tilth refers to the physical condition of a soil in relation to plant growth and it, therefore, embraces all those conditions which promote plant growth.

Starch: Starch is polysaccharide which is the most important form of carbohydrate in the plant. Most of the starches are a mixture of a straight chain polysaccharide called amylose, and a branched chain polysaccharide called amylopectin.

Sub Soil: The layer or horizon of the soil below the surface soil which is generally light in colour, compact and less fertile, is called sub soil.

Sustainable Agriculture: Sustainable agriculture may be defined as that kind of agricultural production, the productivity of which always remains satisfied with the changing of human needs. Moreover, it also assists in enhancing the quality of the environment and conserving the natural resources.

Symbiotic Nitrogen Fixation: Fixation of atmospheric nitrogen by the leguminous plants with the help of root nodule bacteria i.e. the Rhizobia. (Singular Rhizobium) is known as symbiotic nitrogen fixation.

Temperature: The degree of hotness and coldness which is usually expressed either in degrees Celsius or Fahrenheit, is called temperature.

The system of mixed/intercropping renders a very important contribution towards checking the lack of balanced diet of the people. For instance, growing cereals provides carbohydrates, pulses proteins and edible oil seeds fat in the food.

Tillage: The word tillage stands derived from the Anglo-Saxon words tillian and teolin which mean to plough and prepare the soil to sow seeds, to cultivate and raise crops. In other words, it is the mechanical manipulation of the soil to obtain a seedbed of optimum soil tilth for plant growth.

Trap crops: Plant stands which are grown to attract insects or other organisms so that the target crop escapes from the pest attack. For instance, okra can be used as trap crop around cotton for jassid, American bollworm and spotted bollworms. The main benefit of trap crop is that the insecticides are seldom required to be used on the main crop and this enhances the natural control of pests.

Urban or Town Compost: It is prepared from night soil, street and dustbin refuse.

Vermicompost: Compost which is prepared from the digestion and discharge of refuse and other organic wastes by using earthworms such as Eisenia fetida or Eudrillus eugeniae, is termed as vermicompost.

Vesicular Arbuscular Mycorrhizae (VAM): A symbiotic association of fungi with the roots of vesicular plants and has beneficial effect on plant growth particularly of P-deficient soils is known as VAM.

Water Holding Capacity: The moisture saturated stage of the soil when all pores both macro and micro and capillaries are filled with water is known as water holding capacity.

1

Introduction and Need for Organic Farming

Although after independence, the use of hybrid and composite seeds of various crops, especially of rice and wheat resulted in an increase in the Indian agricultural output leading to "Green Revolution". This Green Revolution came to fore during the late "Nineteen Hundred and Sixty". In fact, during the era of "Green Revolution", introduction of high yielding varieties of rice and wheat crops were grown which responded to more use of fertilizers, pesticides as well as water.

There is no doubt that "Green Revolution" made the country-India, self-sufficient in food grains production, particularly of rice and wheat. As a matter of fact, the "Green Revolution" created a buffer stock at times exceeding to 60 million tonnes, which made the nation or the country very proud (Mahajan and Gupta, 2009). However, continuous use of chemical fertilizers as well as inorganic pesticides and totally neglect of organic manures and botanical pesticides prepared from various plant species like Neem (*Azadirachta indica*) and Drek (*Melia azedarach*) for controlling of insects came totally to stand still. So very soon, the deleterious or ill effects of the "Green Revolution" the so called the chemical-based farming or "Industrial agriculture," came to fore, which briefly stand discussed below

Accumulation of nitrate (NO) in underground water: The most deleterious or harmful effects of chemicalized based farming was an accumulation of excessive amount of NO_3 in the underground water which caused an infant disease known as Methemoglobinemia or the Blue baby Disease (Gupta, et al. 2022)

1) *Nitrosamines*: Nitrosamines formed by reacting with excessive amount of nitrate with secondary amines serving as carcinogenic agents.

2) *The excessive nitrate:* The excessive nitrate leached to water bodies along with phosphate increases algal growth in their surface leading to accumulation of considerable of masses dead organic material which in

the long run produces deficiency of oxygen supply and finally influences the aquatic life. This phenomenon is known as "Eutrophication" (Gupta, 2019; Gupta *et al*, 2022).

3) *Accumulation of heavy metals:* Heavy metals which by definition are those elements which have a density greater than 5 in their elemental form. They mainly comprise of some 38 elements. However, the term usually refers to 12 elements that are used and discharged by industries. They mostly consist of Cd, Cr, Co, Cu, Fe, Mn, Hg, Ni, Pb, Sn, Zn. Amongst these elements Cd, Hg, Ni, Pb, Cr, Co, Cu and Zn may represent potential hazards both to plants and animals. Accumulation of heavy metals like Cd, Pb and Cr in soils and water, are the principal causes of environmental concern due to an excessive use of fertilizers. Further it is added that excessive use of nitrogenous fertilizers causes more production of gases in the form of nitric oxide (NO), nitrogen dioxide (NO_2) and dinitrogen oxide (N_2O) which adversely affect the ozone layer.

4) *Effects of Pesticides:* Like chemical fertilizers, the use of pesticides was started with the introduction of high yielding varieties of various crops. In fact, the new introduced varieties of different crops like rice, wheat, sugarcane, oil seeds, millets and many others possessed more response to applied nitrogenous, phosphatic and potassic fertilizers and as a result they are liable to become more susceptible to an assault of different insect pests and diseases. Crop loss due to pest's attack was found to range from 10-30 per cent depending up on the crop and environment. Annual crop loss owing to pest's attack in India was estimated to about Rs. 200000 million per year. Thus, use of pesticides has now become an increasingly necessary operation for controlling pests. Hitherto, pesticides were used in small quantities, thus without releasing any adverse and toxic effects on the agricultural ecosystem. Now, more but indiscriminate use of pesticides has however, become a very serious matter because of health hazards to human beings, animals and plants vis-à-vis their effect on wildlife (Gupta, 2006, Gupta, 2016). Pesticidal pollution is, thus, becoming a major environmental pollution and its abatement must receive attention to all.

Large amounts of pesticides are sprayed on various crops and fruit trees as well as vegetables (Bhat 2021). These get washed into small rivulets and streams and end up in the rivers tube wells and wells. If there is rainfall after the pesticides spray the drinking water sources get more contaminated. There is also evidence that many of these pesticides have brough alteration in the soil biological ecosystem like bacteria actinomycetes, fungi, algae as well as protozoa, nematodes, earthworms, millipedes, spiders, slugs, snails, mites, ticks, spiders, insects etc. Many studies have indicated that pesticide exposure is associated with the long-term health problems such as respiratory problems, memory disorders, dermatologic condition, cancer problem, depression, neurological deficits, birth defects and miscarriages (Priyadarshini et al. 2013).

Fluoride ions have both the negative and positive impact on human health. It is considered beneficial for human health if it is taken in controlled quantity i.e., 1.5 to 1.5 mg L^{-1} of water. But more than this figure, it then produces dental fluorosis.

Thus, the present "Agriculture" is in deep crisis both in human beings and ecological variables. Two dimensions of the human's crises are the suicides of the farmers and the growth of hunger and malnutrition. The agrarian crises leading to farmers suicides is a result of debt, and debt is a result of the convergence of rising costs of non-sustainable and in appropriate production systems and falling prices of agricultural products due to unjust and unfair trade patterns (Shiva, 2007).

While the farmers were dying due to debt and negative incomes, the poor were being denied their right to food. It is point to mention that about 30 % rural households in India were eating 1,600 kcal in 1998 compared to 1820 kcal in 1999 (Shiva, 2007). In 1999-2000, almost 77% rural population consumed less than the poverty line calorie requirement of 2,400 kcal. One third of all hungry children in the world are today in India.

The ecological crises engineered by the industrial chemical farming include:

1) Toxic contamination
2) Biodiversity erosion
3) Water depletion and its heavy metal pollution.
4) Desertification of soils
5) Pollution of the atmosphere because of fossil fuels, which is leading to climate change.

It is remarkable to note that climate change rapidly evolves into climate catastrophe if greenhouse gas emissions are not reduced. Industrial agriculture with its chemical inputs (chemical fertilizers and pesticides) is responsible for 25 per cent of the world's carbon dioxide emission, 60 per cent of methane gas emissions and 80 per cent of nitrous oxide, all these gas emissions are powerful greenhouse gases. It is point to mention that nitrous oxide is 200 times more potent than carbon dioxide as a greenhouse gas and is mainly produced by the use of nitrogenous fertilizers.

Emissions of carbon from the burning of fossil fuels for agricultural purposes in England and Germany were as much as 0.046 and 0.053 tonnes per ha, while they are only 0.007 i.e., roughly seven times lower in non-industrial agriculture.

Industrial agriculture uses ten times more energy as input than it produces as calories for food on the other hand, biodiverse ecological organic agriculture absorbs carbon dioxide and puts it into the soil as an organic matter and humus (well decomposed organic matter). Humus also helps to reduce the impact, of droughts and floods, which are becoming more frequent and intense as a result of climate change. Biodiverse organic farming, thus, contributes to mitigation of and adaptation to climate change resulting from fossil fuels and emission of greenhouse gases.

Credit for "Organic Farming" on global scale goes to Japanese Plant Pathologist Masanobu Fukuoka who secured FAO award for the signal service made to the cause of "Organic Farming". He advocated that soil's nourishing qualities stem from organic matter which returns to the soil periodically. Fukuoka took about 25 years to develop rice variety known as "Fukuoka rice", was the best genetically modified by the natural breeding using only organics. This rice variety has given promising yield in the coldest places of all over the world. Later on, the European Union had passed strict regulation on the organic produce and the farmers have to observe the same as an obligation to label their produce as "Organic Produce". "Organic farming thus refers to farming in the spirit of organic relationship (Partap, 2009). It is such a productive system which always shuns excessive use of chemical fertilizers, pesticides, growth regulators and livestock additives. Organic Farming system mainly relies up on the use of organic manures i.e., farmyard manure, compost, green manuring, botanical pesticides, biocides and cultural methods. Organic farming, however, does not mean reverting to "Stone Age Agriculture" (Gupta, 2023).

Considering the harmful effects of chemical fertilizers, pesticides and other agrochemicals on the health of humans, animals, wildlife, water pollution as well as soil and air pollutions i.e., environmental degradation as stated above. Organic Farming needs to be promoted not only in India, but in whole of the world. Other names of organic farming also stand stated: -

Other Names of Organic Farming

Rishikrishi Agriculture: In India Acharya Vinoba Bhave conducted an experiment at his Paunar Ashram (Wardha) to till the land with the bare hands as our ancient Rishis (Hermits sages who wrote the Vedas) used to do for meeting their food and fiber requirement. This is called the "Rishi Krishi or

Do-Nothing Agriculture Approach, this approach, however did not run very well on large scale bases to feed the burgeoning human population as has been happened today (Chhonkar, 2003).

Natural Farming: It is a system of agricultural production devised by Masanobu Fukuoka (Japanese Plant Pathologist) that seeks to follow nature by minimizing human interference i.e., no mechanical cultivation, no weeding by tillage or herbicide, no dependence on use of chemical pesticides, no use of synthetic fertilizers.

Ecological Agriculture or Eco-farming: It is a type of farming where under such practices are followed that enhance or at least do not harm the environment and are aimed at minimizing the use of chemical inputs, rather than completely avoiding them.

Sustainable Aagriculture: A system of farming whereunder all management resources of Agriculture are resorted to satisfy changing human needs as well as maintaining or enhancing the quality of environment and conserving the natural resources.

Homa Organic Farming: This is the knowledge relating to farming which can be utilized for agricultural production without use of chemical fertilizers and pesticides. This is HOMA THERAPY farming. The process of AGNIHOTRA PYRAMID FIRE is the basic HOMA to sustain the life.

Agnihotra and Microbes: Agnihotra organic process is a metabolic process of plants in which fat (Ghee) is used as a catalytic agent for catalytic factor.

Biodynamic Farming: Organic farming or organic agriculture is also called by another name i.e., Biodynamic Farming which stands described separately i.e., Biodynamic Farming and its scope in India.

It is concluded from this chapter of the book that now a days "Organic Farming" is gaining gradual momentum across India and the world. Growing awareness of health and environmental issues has demanded production of organic food. While trends of rising consumers demand for organic food. While trends of rising consumer's demand for organic are becoming discernible, sustainability in production of food grain crops, fruit trees and vegetables has become the prime concern in agriculture, horticulture and vegetables development. As a matter of fact, organic foods are getting a niche market in both domestic and export market with these opportunities, there are different schools of thought, differing mainly on the challenge of food security and feeding the teeming of millions.

It is worthwhile to mention that organic agriculture is not a name of group of technologies or a movement or approach to make agriculture near to nature, simple, sustainable and safe to the society to get nature's gift. With the research in last 38 years and experience gained world over, it has been found that the "Organic Farming" is a safe approach for solving environmental problems. Moreover, it has great to make self-sustainable life in villages, also with safe food produced from organic farming, the consumers will be less prone to health problems. Organic agriculture is not just exclusion of synthetic chemicals from production system. It needs to be viewed in broader perspective of solving environmental problems as pointed out by Sharma (2009). According to Lal (2022) organic farming is a method of farming system which primarily aimed at cultivating the land and raising crops in such a way, so as to keep the soil healthy by way of using organic wastes (well decomposed ones) and other biological methods. As a matter of fact, by performing such activities plant nutrients are provided to the growing crops which as a result enhance sustainable productivity in an eco-friendly and pollution free environment (Gupta et al. 2022). Similarly, biopesticides can be used because they are also eco-friendly biodegradable, non-toxic and low cost effective alternative for chemical pesticides to increase the crop yield on sustainable basis.

Although organic farming is expensive, labour consuming and is also restrained by a number of constraints for availability of organic matter, transport from source to site and lack of price premium yet in the long run it has proved beneficial not only to the farming communities but also to the environment both abiotic (land, air, water) and biotic (animals and plants) Moreover, organic farming has proved useful in maintaining the sustainability in agriculture production(Gupta, 2023) Moreover, the transformation from subsistence to sustainable agriculture in the Union Territory of Jammu and Kashmir has to come up through investment of knowledge, technology and capital (Kumar, 2022). Hopefully, the committee with members having vast field experience and intellectual steer the agriculture sector in the aforesaid Union Territory to a new both of sustainability remunerativeness and empowerment as reported by Kumar (2022).

Use of organic inputs in India: In India, use of organic inputs for food production has been old age practice (Gupta, 2019). For instance, use of dung as organic manure appears to have been practiced since the Rigvedic Age (2500-1500BC). Similarly, the magnitude of green manure has been known as far back as 1000BC, as reference to use of stalks and stems of sesame as manure has been known as back as far 1000 BC (Gupta, 2019). Farmers of union territory of Ladakh are following traditional method of farming.

Due to indiscriminate use of chemical fertilizers and pesticides as stated in Chapter 1, there is a growing concern about the various environmental and health hazards along with socio economic problems which are non-sustainable. Moreover, these are not in consonance with economics and ecology. Although due to application of chemical fertilizers and pesticides, the agricultural production increased initially but the productivity per unit area has started to decline over the years (Gogoi, 2015). This has led many of the peasants to leave agriculture profession and many of them committed suicides. However, some of the consumers to seek alternative practices and systems which will make or render agriculture profession more sustainable. Hence, there is, in fact, the time to come back to our own rich agricultural practices or opt for an alternative sustainable agriculture practice which are environmentally and ecologically sound, socially more acceptable and economically more profitable.

Alternatives Among Chemical Agriculture

Among the various alternatives biodynamic farming or biodynamic agriculture is found to be the most suitable and the sound alternative in this aspect. Biodynamic agriculture is an advanced form of organic agriculture with an emphasis on food grains quality and soil productivity and as such no synthetic chemicals (fertilizers and pesticides) is used. As is evident "Biodynamic" is derived from two "Greek Words", "Bios" means life or living and "Dynamos" means energy that refers to the agricultural science that recognizes the basic principles of work in nature and applies this knowledge of life forces to bring back balance and healing in the soil. It revitalizes the biological as well as the chemical values in soil which are rapidly becoming depleted through modern agriculture techniques. In other words, living soil with increased biological activity in alignment with planetary cosmic rhythms particularly those of sun, moon and planets in the fixed constellations. The coordinated working together of earthly and cosmic energies is what brings about healthy and nutritious plants.

Base of Biodynamic Agriculture

Biodynamic agriculture is based upon Dr. Rudolf Steiner's Philosophy. Anthroposophy Anthroposophical ideas have been applied in a range of fields including agriculture. The methods are based on a series of 8 lectures known as the, "Agriculture Course." His idea was influenced to a great extent by oriental philosophy specially, Buddhism, Hinduism and the Vedic scriptures. He showed how the health of the soil, plants and animals depend up on reconnecting nature with creative forces of cosmos. As a matter of fact, it is a complete holistic approach or outlook on the agriculture profession and an advanced form of organic farming with techniques to farm the air as well as farm soil and is the oldest organic farming movement practised in more than 20 countries of the world. It includes the normal organic farming practice, consisting of the use of compost, farmyard manure, green manure and crop rotation.

Apart from the above, biodynamic farming uses a series of preparations numbered from BD 500 to BD 508 and CPP. BD 500 (cow horn manure which is basically fermented cow dung. It is basis for soil fertility and renewal of degraded soil. It is burned in September to November and lifted from the field during February/March. Contrary to this, BD-508, is Horsetail Herb-*Equisetum arvense* used for soil on foliar application, when conditions are wet helps to discourage fungal diseases, CPP i.e., Cow Pat Pit or Pit Pat is another preparation which is also used in minute quantities. All these preparations are based on various mineral, plant and animal substances. These enhance all the bacterial, fungal and mineral processes that are found in organic farming, system, placing great importance on the auspicious positions of the moon, sun and planets calendar is utilized for applying the biodynamic farming preparations, sowing seeds, planting plants, applying liquid manures, spraying fruit trees.

They are used in other farming activities to control weed, pest and rodents. This shows remarkable effects on growth, metabolism, crop yield and quality, thereby, capable of affording long term sustainability to agricultural ecosystem.

Biodynamic Movement in India

Biodynamic farming started in India in early 1990 with the help of a soil scientist, Peter Procter from New Zealand working in biodynamic farming since 1965, from an early initiative started at the Kasturbagram Ashram, Indore. 1t has spread like wildfire from Nainital to Kodaikanal and from Mumbai to Orissa. The farmers have been able to produce biodynamic coffee, cotton, and vegetables with biodynamic produce certification by DEMETER. In India, Biodynamic Association of India (BDAI) came to be established with its headquarter at Bangalore.

Organic Farming (Biodynamic farming)- History, Benefits and its Scope in India 11 Organic Farming (Biodynamic farming), History Benefits and Its Scope In India .

The primary objective of BDAI was to promote and coordinate biodynamic movement in India. However, interest in Biodynamic farming has greatly increased over the past 10 years i.e., with effect from among organic farmers in India. Up to January 2015,there were more than 500 small and big farmers practicing Biodynamic agriculture throughout the country (Gogoi, 2015) Among the first initiatives were the Kurin ji farms producing a wide range of crops including fruits vegetables (cabbage, broccoli, lettuce coriander etc) all pulses, cucumber and tomatoes are among the many quality varieties grown in Bhaikaka Krishi Kendras at Anand Gujarat. Fruits such as papaya, limes including mosambi are growing well. There are also Biodynamic apple orchards at Nainital. Biodynamic rice at ISKON farm (Ahmedabad) as well as biodynamic, sugarcane, vanilla, tea and coffee plantations are in South India.

Biodynamic farming though possesses a lot of ecological implications yet it is a form of non-chemical and non-toxic farming. There is a greater need to expand and encourage these alternatives form of agriculture that uses no synthetic or chemical inputs. The farmers can become self-sufficient by adopting this kind of organic farming without chemical-based fertilizers and pesticides. They can achieve a higher level of productivity and brings back vitality into the food that we consume. As north east India has wide potential for organic farming so there is also vast scope to adopt biodynamic farming by the farmers especially in North West Himalayan Region of the country. However, lot of research is required to be resorted for its feasibility in this region as well as others.

References

Bhat, Raja Muzaffar. 2021. Pesticides in our drinking water. The Daily Excelsior. 57(256):8

Chhonkar, P.K. 2003. Organic farming—Science or belief. J.Indian Soc.Soil Sci.51:365-377

Dhaliwal, S. 2015. Punjab facing a veritable water crisis. The Tribune 135 (234):9 [25th August Tuesday]

Gogoi, D. 2015 Biodynamic Agriculture: A holistic way of farming Agrobios Newsletter. 13 (8): 42-43.

Gupta, R. D. 2006. Environmental Pollution: Hazards and Control, Concept Publishing Company, New Delhi - 110059.

Gupta,R.D.2011.Traditional farming. The Daily Excelsior.47(109):6

Gupta, R.D. 2016. Wildlife and wetland Ecosystem of Jammu and Kashmir Himalayas Published by N.R. Book International, Pacca Danga Jammu Tawi, J&K, India.

Gupta, R.D.2019. Organic farming: Need of the day. The State Times 24 (5): 6.

Gupta, R.D. 2023. Organic farming: A form of sustainable agriculture. The State Times. 28(38):6 Gupta, R.D., Gupta, S.K. and Mahajan A. 2022. Sustainable System of Farming for Modern Agriculture. Nipa GENX Electronic Resources P. Ltd New Delhi-10034, India.

Gupta, R.D. Gupta, S.K. and Mahajan, A. 2022. Sustainable Systems of Farming for Modern Agriculture, Nippa Genx Electronic Resource and Solutions, Pvt. Ltd. New India Publishing Agency, New Delhi.

Gupta,R.K and Gupta,R.D. 2004. Ecofriendly biopesticides. The Daily Excelsior.40(341):6

Kumar, Parveen, 2022. Toward holistic development of agriculture in J&K The State Times. 27(252):6.

Lal Banarsi, 2022. Promoting Organic Farming in J&K. The State Times. 27(90):6.

Mahajan, A and Gupta R.D. 2009. Integrated nutrient management (INM) in a sustainable rice-wheat cropping system, Springer Science Business Media (www.springer.com).

Partap Tej. 2009. Organic elixir- "A feasible option, Agriculture Today,12 (6): 28.

Priyadarshini, S.R, Manjula, B., Umadevi, M. and Amutha, C.2013. Panchagavya: The Organic way of crop protection. Agrobios Newsletter 11(11): 27-29.

Reen, A.S. 2003. Organic Farming for Sustainable Agriculture. The Daily Excelsior 39(56):8.

Sharma, A.K. 2009, Organic Agriculture Agriculture Today. 12(6):29.

Shiva, Vandana. 2007. We need to go organic: Responding to the agrarian crisis and climate. chaos. Farmers Forum. 7(1):4-5.

2

History and Importance of Organic Farming

History

The history of organic farming is largely based upon the history of organic movement which came to fore as a reaction against the large-scale use of chemical fertilizers and pesticides in agriculture. This movement can be traced back to England during 1920, when individuals started speaking against a variety of agricultural "innovation".

One of the earlier pioneers of Organic Farming, Sir Albert Howard, is often referred to as the father of the modern "Organic Farming". He began to work as a British botanist and served as an agriculture advisor in India, where he observed traditional Indian farming practices and came to respect them as superior to his conventional agriculture science. He documented and developed these organic farming practices/methods. His writings and notably the 1940 book, An Agricultural Testament, have influenced many scientists and farmers of the day.

In Germany Dr. Rudolf Steiner developed biodynamic agriculture which was probably the first comprehensive farming system. In America, the agronomist F.H. King had toured China, Korea and Japan in early 1990s, closely studied the fertilization, tillage and general farming practices being followed there. He published his findings in his book Farmers of Forty Centuries (1911). King probably did not impart any view himself as part of the movement, whether Organic Farming movement or by some other name. However, in the later years, some scientists within the organic farming movement thought, about what King had encountered in Asia as kindred approach.

During 1930 strongly influenced by Sir Howard's work, Lady Eve Balfour launched the Haughley Experiment on farmland in England. It was the first and the foremost conventional farming side by side the comparison of organic farming. After four years later, she published the living soil based on the initial findings of the Haughley's Experiment. It was widely gone through and led to the formation of a key International Organic Advocacy Group, the Soil Association.

The first wide usage of the term organic farming is usually credited to appear to Lord North Bourn in his book known as "Look to the land (1940)," where under he described a holistic approach to the ecological balanced farming. Later on, a well-known Western Scientist thought to study and developed traditional methods, while most of the research and development concentrated on the new chemicalized approach. This fundamental split in how science viewed agriculture was mainly responsible for the organic versus chemicalized farming situation of today.

In a parallel development in Japan, Masanobu Fukuoka, a microbiologist working in soil science and plant pathology began to have doubt to the modern agricultural movement i.e., on Green Revolution. In the early 1940s, he left his job as a research scientist, returned to his family's farming and devoted about 30 next years in developing a "Radical No Till Organic Method" for growing food grains, which is now known as "Fukuoka Farming" or "Fukuoka Organic Farming."

During the World War II, Technological Information Relating to Organic Farming.

Importance

Organic Farming or Organic Agriculture is of great economic importance, especially these days, when the "Green Revolution," in the country has become fatigued (Mahajan and Gupta, 2009). As a result, it has either started a constant production of rice and wheat crops year after year or their yields have decreased considerably. It is mainly attributed to use of chemical fertilizers and pesticides in these crops year after without any use of organic manures and organic pesticides. The chemicalized intensive agriculture has, in fact, rendered a small farmer into marginal and marginal into landless." Not only this, the chemicalized and pesticides-oriented cotton cultivation has not merely ruined the entire ecosystem and contaminated the environment, but it also crippled the health environmental equilibrium of the people dwelling in the Malwa area. As many as 3,000 farmers committed suicide only in this region. Most of them who committed suicides were cotton growers who faced the rapid failure of cotton crop in nineties. As such hundreds of them shifted to paddy cultivation, pushing already water scarce Malwa in crises of ground water over exploitation (Anonymous 2005).

Organic farming or organic agriculture is a system of farming based upon integrated relationship among the soil, water, plant, macro and microorganisms vis-a-vis farm animals. Thus, it creates a productive landscape and successfully reconciles food production and environmental conservation (Arora, 2014). Organic farming management relies up on local human resources and knowledges to enhance natural resource processes, respecting ecological carrying capacities. By reducing dependence on off farm inputs and creating more balanced nutrients and energy flows ecosystem resilience is strengthened, food security is increased and additional incomes are generated. It is point to mention that organic farming responds positively to all sustainable agriculture and rural development objectives and assists in sustaining fertility of soil and thereby crop productivity.

Careful planning of crop associations and good planning of animal husbandry will help to build up lot of organic manures (FYM and Compost). Adding of these organic manures promote soil aggregation, check soil erosion and increase water retention capacity of the soils as well enhance the rate of water percolation. This organic farming plays a significant role in soil and water conservation systems of successful organic farming increase and encourage biodiversity through multiple cropping pattern and crop rotations. Integrated crop tree animal system, the use of traditional and adapted food and fodder species and creation of wildlife habitats to abstract pest enemies and pollinators.

Organic agriculture encourages farmers to manage locally available resources, reducing dependence on external inputs or food distribution systems over which they have little or no control. Further, participatory research has a great potential to ameliorate organic food production over conventional yield levels, especially in area with limited access to conventional inputs. Organic farming improves food security by way of increasing income opportunities. Through polycultures, raised fields and agroforestry systems, organic farmers sustain year-round yields, reducing chances of crops failures. This organic agriculture is very important to the food security particularly of poor peasants and farmers living in fragile or market marginalized areas.

Good agriculture management is vital to success of organic operation as the farmers cannot resort to synthetic inputs, if problems emerge. Practices of biological pest and disease control minimize their incidents, hormones and antibiotic causing health hazards to farmers. Organic farming has a large number of benefits. In the first place, organic foods have extremely good fragrance and tastes. This is the message which has conveyed by many people who eat organic foods. Fruits and vegetables grown organically, have shown more content of nutrients, vitamins, minerals, and cancer fighting antioxidants than non-organic grown fruits and vegetables. Organic food production is more sustainable and friendly to the environment both to physical (land, air and water) and biological (plants and animals). There is no air,water and soil pollution. Organic farming system improves soil's physical, chemical and biological properties.

In rice wheat organic farming system, soil microbial population (bacteria. actinomycetes, fungi and blue green algae) was enhanced due to the application of organic amendments in comparison to recommended fertilizer application (Sharma, 2014). This also increased the amount of soil organic carbon and available phosphorus contents. Consumption of organic foods by the babies makes them free from blue baby disease which has been created due to the consumption of water contaminated with nitrate (NO3) or foods having NO3 in them.

It is concluded that in late 1970's farmers world over, realized the adverse effects of conventional farming and started efforts to develop a sustainable system in collaboration with the aids of agricultural scientists. During 1990's consumers also realized the ill effects of chemical fertilizers and pesticides. This ultimately instigated the demand for safe and sustainable food production that forced policy makers to promote Organic Farming or Organic Agriculture.

India has recorded considerable development of organic farming. In 1999, only 40,000 ha area was recorded as certified organic which increased up to 2,40000 ha during 2011 (Sharma 2014) within a decade. Now, India is a home to 30 percent of the total organic producers in the world but accounts for just 2.59 percent (1.5 million hectare) of the total organic cultivation of 57.8 million hectares according the world of Organic Agriculture 2018 report Dec 10(2018). India has the highest number of organic farmers globally but most of them are struggling. Poor policy measures rising input costs and limited market are affecting growth of organic farming in the country.

First Use of Organic Farming is Germany

Although in Germany, Rudolf Steiner's Spiritual Foundations for the renewal of Agriculture published in 1924 led to the popularization of biodynamic agriculture, one of the first organic farming systems in 1939, yet the first use of the term "Organic Farming" is usually credited to Lord Northbourne, in his book "Look to the Land" (1940), wherein he described a holistic, ecologically balanced approach to farming (Singh and Singh, 2007). Apart from these aspects in 1939, Lady Eve Balfour influenced by Sir Howard's work, launched the first scientific side by side comparison of organic and conventional farming in the England, called the Houghtley Experiment. It was documented by

Lady Balfour in her book, "The Living Soil (1943)". Its influence led to the formation of the "Soil Association, during 1946; a key international "Organic Advocacy Group"

The first use of the term organic farming is usually credited to Lord Northbourne, in his book "Look to the Land (1940) wherein he described a holistic, ecologically balanced approach farming".

Organic Farming in Central Europe, India and United States

While some indigenous cultures had been farming organically for centuries, organic agriculture began to develop consciously in Central Europe and India during the early twentieth century as a reaction of insiders (Agricultural Scientists and Farmers) against the "Industrialized Agriculture", brought about through the "Green Revolution" (Singh and Singh, 2007). The British botanist, Sir Albert Howard often called, "The Father of Modern Organic Agriculture", studied traditional farming practices in west Bengal, India. He came to regard such practices as superior to "Modern Agricultural Sciences i.e., "Green Revolutionized Agriculture" and recorded them in his book which was named as, "An Agricultural Testament".

During the 1950's, sustainable agriculture became a topic of scientific interest although research focused on developing the new chemical approaches. In the US, J.I. Rodale popularised organic gardening among consumers. In the 1970s, global movements concerning with the environmental championed organic farming. As the distinction between organic and conventional food became clearer, one goal of the organic movement was to encourage consumption of locally grown food, which was promoted through slogans like, "Know Your Farmer, Know Your Food". In 1972, the International Federation of Organic Agriculture Movements (IFOAM) was founded in Versailles France. IFOAM was dedicated to diffusion of information on the principles and practices of Organic Agriculture across national and linguistic boundaries.

In the 1980s, various farming and consumer groups worldwide started pressing for government regulation of "Organic Production." This led to legislation and the certification on standards being enacted beginning in the 1990's.

Since the early 1990's, the retail market for organic farming in developed economy have grown about 20 per cent annually due to increasing consumer's demand. While small independent producers and consumers initially drove the rise of organic farming, increasingly organic market growth has led to the participation of agribusiness interests. As the volume and variety of organic products grows, production is increasingly on large scale.

Processes or Methods of Organic Farming

Organic farming involves fostering of natural processes, often over extend periods of time and a holistic approach. A variety of methods are employed.

They mainly consist of crop rotation, cover cropping, green manuring, mulching application of farm yard manure, compost including vermicompost. Organic farmers also use processed natural or organic fertilizers such as seed meal, neem cake, bone meal dried blood, molasses, oil cakes (mustard and rape seed cakes, castor cake, cotton seed cake etc).

Biological Control

The biological control, being ecofriendly helps in reducing the dependency on chemical fungicides. In recent times, biological control has attracted the attention in managing the disease problems. Species of *Tricoderma* and *Gliocladium* have been extensively exploited for their utility as bioagents against several soil and seed borne pathogens of various crops. In wheat crop such efforts were concentrated towards the control of seed borne diseases, loose smut and Karnal bunt. Being a part of soil plant system, species of *Trichoderma* were found among the most desired ones.

Use of Ash

Sprinkling or sprawling of ash dust in growing vegetables belonging to Cucurbitaceae family like gourd, pumpkin as well as watermelons, muskmelons etc, is made. It is done to repel the insect pests of these vegetables.

Use of Neem Tree (*Azadirachta indica*)

The leaves of neem tree are dried. After drying, these are wrapped in the paper. The wrapped papers are then put in top, middle and at the bottom of the granaries. Later on, the granaries are closed air tightly. This technique helps to save the food grains from the attack of the insect pests.

The farmers of Akhnoor Tehsil of district (Jammu and Kashmir Union Territory) were previously used the above said practice. But with the Green Revolution, the use of insecticides, fungicides and other chemicals became necessary. As such, the indigenous technique as given above, came into halt (Gupta, 2019 a, 2019 b).

Use of Cow's Urine

Many of the traditional farmers still use the urine of cows to combat crop pests and diseases. When cows' urine is sprayed at 10 percent concentration to diseased plants, they soon recover and become healthy.

Crop Diversity

Crop diversity is also characteristic of organic farming. Planting a variety of vegetable crops supports a wider range of beneficial insects, soil microorganisms and other factors that add up to overall farm health but requires close attention.

Raising of Livestock and Poultry under Natural Conditions

Organic farms that raise livestock and poultry for meat, dairy products and animals with "Natural" living conditions and feed. Ample, free range outdoor access for grazing and exercise is a distinctive feature and crowding is avoided. Fodder available is also naturally grown and drugs including antibiotics are prohibited. Animal health and food quality are thus pursued in a "fresh air, exercise and good food approach (Singh and Singh, 2007).

Organic Farming System

There are a number of organic farming systems which prescribe specific techniques. In fact, they tend to be more specific and can always be fitted within general organic farming standards. These are: i) Biodynamic Farming and ii) Natural Farming which already stand mentioned.

The former is a comprehensive international approach with its own international governing body. The latter is a no-till system for small scale food grain production. French Intensive and Bio-intensive systems that go beyond organic principles and approach sustainability.

Organic Farming System in India

The concept of organic farming is not new but it is being followed from ancient time or time immemorial. It is a method of farming system which primarily aimed at cultivating the land for raising crops in such a way, as to keep the soil alive and in good health by use of well-prepared organic wastes (crop, animal, farm and aquatic wastes) via-a-vis other biological materials. Such biological materials consist of organic manures like farm yard manure (FYM) compost, vermicompost as well as biofertilizers prepared from beneficial microbes to release essential nutrients for growing crops to obtain sustainable agricultural production. At the same time, organic manures are ecofriendly and pollution free environment.

As per the definition of the USDA study team on organic farming "Organic farming is a system which avoids or largely excludes the use of synthetic inputs (such as fertilizers, pesticides, hormones, feed additives etc.) and to the maximum extent feasible rely up on crop rotations, crop residues, organic manures, off-farm organic wastes, mineral grade rock additives and biological system of mobilization and plant protection".

In another definitions, FAO suggested that "Organic agriculture is a unique production management system which promotes and enhances Argo ecosystem's health, including biodiversity, biological cycles and soil biological activity, and this is accomplished by using on - farm agronomic, biological and mechanical methods in exclusion of all synthetic off-farm Inputs".

Need of Organic Farming

Due to increase in human population which is presently 135 crore's (July, August, 2020), our compulsion would be not only to stabilize agricultural production but to increase it further in sustainable manner. Agricultural scientists have realized that the "Green Revolution" with high input use (chemical fertilizers and pesticides), has reached a plateau and now sustained with diminishing return of falling dividends. As a matter of fact, the yield per hectare of various crops, especially rice and wheat which marked a significant increase in the "Green Revolution" days, has not shown further increase of these crops rather their yield decreased considerably. Rising input costs and low returns to farmers became them indebtedness. The policy makers believed that loan against loan could bail out the farmers. But this led to accumulation of loan and increased in burden for repayment, particularly in years when the farmers did not get good returns for their produces after spending heavily on input costs. Thus, indebtedness was the ultimate cause for farmers distress and suicides (Sharma, 2014). Majority of the farmers who are driven to suicide borrow from private money lenders including commission agents as well as big farmers, mostly under emergency situations of exorbitant rates of interest. They were not able to repay these loans because there were invariably used for nonproductive agricultural purposes. When these borrowers were pressed for repayment, they were often compelled to resort bank loans. They take crop loan or borrow further for buying tractors, which they immediately sold in the market at lower price and retired the debt obtained from private money lenders. In some cases, the cheap rate credit facility was misused. Some big landholders have been borrowed from the banks to the limit of their entitlements even they do not need that money. This low-cost money they lend to the poor farmers at higher rates of interest like many lenders.

Some examples of self-inflicted irrationalities are provided in a report submitted by a committee on farmers distress set up by the Reserve Bank of India under the chairmanship of S.S Johl where the suicides/deaths were not even remotely related to the farmer's indebtedness and yet were treated as farmers suicides. In spite of the above quoted fague example by S.S. Johl, still the fact remains that debt is one of the major causes of farmers suicides and deserves serious handling by the policy makers. Surveys and resurveys repeated many times do not offer a solution. Even a single suicide by a farmer due to indebtedness should be a matter of serious concern. Also paying the compensation to the affected families after the suicides is only a palliative that does not eliminate the courage of suicides. It is just the preventive measures which are the need of the hour. Apart from the above the recent findings of high limits of pesticide residues in beverages like Pepsi and Coca cola had initiated a debate in the country and went to the extent of banning the sale and production. Ban on these two products, however, was not a solution to this problem. The Centre for Science and Environment was asked to conduct the study in this connection and accordingly the norms were formulated.

Because of indiscriminate use of chemical fertilizer, after "Green Revolution" for decades, the organic matter content of soils has come down to less than 1 percent. In addition, the use of pesticides led to the pest resurgence and difficult to control weeds. The residues of the chemicals are causing serious concern over the safely of food and sustainable production as well as Argo ecosystem. Indiscriminate consumption of nitrogenous fertilizer, especially urea became a potent source of groundwater contamination. It was due to its high solubility and no polarity, and very susceptible to leaching and percolating into the ground water in the form of nitrate (NO_3) produced in the soil, thereby polluting ground water. Heavy irrigations or rains can carry NO_3 ions deep into the ground water. Drinking of NO_3 contaminated water causes "Methemoglobinemia disease in infants which is commonly known as blue baby disease. Hither to underground water having more than 45 ppm of NO_3 used to cause this disease but now its concentration even up to 10ppm has led numerous such cases (Gupta and Singh, 2006).

Beside presence of NO_3 accumulation of heavy metals such as Cd, Cr Ni and Pb have also accumulated in soils over a period of time to toxic level due to indiscriminate use of fertilizer. These metals finally get accumulated in human bodies through consumption of agricultural products grown on such soils.

Above all, application of nitrogenous fertilizers alone without using phosphatic and potassic fertilizers, renders to form some specific kinds of compounds which are suspected to serve as carcinogenic agents. Such compounds named as nitrosamines are formed by the reaction of secondary amines (Gupta and Singh, 2006).

In the light of the above factors or reasons it becomes an imperative to think over the use of organics by adopting "Organic Agriculture" or "Organic Farming". Organic farming is, thus, a production system which helps to produce chemical free nutritious food of high quality. In organic farming system, the soil is always considered as a living system, not an inert bowl for unloading chemicals.

References

Anonymous, 2005. Organic Farming to Boost Punjab's Kharif season. Farmers Forum 2(6):9

Arora, S. 2014. Importance of agriculture. Soil and water conservation Today 9 (4).

Gupta, R.D. 2019a Organic Farming Need of the Day. The State Times. 24(5):6.

Gupta, R.D and Singh, H.2006 Indiscriminate use of poses health hazards. Farmers Forum. 6(9): 20-24.

Gupta R.D. 2019b. Organic Farming Need of the Day. The State Times. 24(6): 6.

Mahajan, A and Gupta, R.D.2009. Integrated Nutrient Management in Sustainable Rice-Wheat Cropping System. Springer Science + Business Media Biv 2009.

Sharma, V. 2014. The socioeconomic aspects of land degradation of lower Siwaliks of Jammu division. Soil and water conservation Today 9(4):5.

Singh, G.P. and Singh, J.2007. Organic Farming in India: - An overview. Indian Farmers Digest. 40(4): 9-14.

3

Organic Farming in India

Historical Perspective

Organic Farming was initiated in about 10,000 years back when ancient farmers started cultivation depending on natural sources only. There is a brief mention of several organic inputs in our ancient literature like Rigveda, Ramayana, Mahabharata, Kautilya Arthashastra etc. In fact, organic agriculture has its roots in traditional agricultural practices that evolved in countless villages and farming communities over the millennium.

Historical Perspective of Organic Farming	
Ancient Period	**Mention of Organic Practices**
Oldest Practice	10,000 years old dating back to Neolithic age practised by ancient cultivation like Mesopotamia, Hwang Ho basin etc
Ramayana	All dead things-rotting corpse or stinking garbage returned to earth are transformed into wholesome things that nourish life. Such is the alchemy of mother earth as interpreted by C. Rajagopalachari
Mahabharata (5500 BC)	Mention of Kamadhenu, the celestial cow and its role on human life and soil fertility.
Kautilya Artha shastra(200 BC)	Mentioned several manures like oil cake, excreta of animals
Brihad Sanhita (byVarahmihir)	Described how to choose manures for different crops and the methods of manuring.
Rigveda (2500-1500 BC)	Mention of organic manure in Rig Veda (2500-1500 BC). Green Manure in Atharva Veda (1000 BC). In Sukra it is stated that to cause healthy growth the plant should be nourished by dungs of goat, sheep and cow as well as water
Holy Quran 590 AD	At least one third of what you take out from soils must be returned to it implying recycling or post-harvest residue.

Organic Movement

The organic movement began in the 1930s and 1940s as a reaction to agriculture's growing reliance on synthetic fertilizers. Artificial fertilizers had been created during the 18th century, initially with superphosphates and then ammonia derived fertilizers mass-produced using the Haber-Bosch process developed during World War I. These early fertilizers were cheap, powerful, and easy to transport in bulk. The 1940s has been referred to as the 'pesticide era'. Sir Albert Howard is widely considered to be the father of organic farming. Rudolf Steiner, an Austrian philosopher, made important strides in the earliest organic theory with his biodynamic agriculture. More work was done by J.I. Rodale in the United States, Lady Eve Balfour in the United Kingdom, and many others across the world. As a percentage of total agricultural output, organic farming has remained tiny since its beginning. As environmental awareness and concern increased, the originally supply-driven movement became demand driven. Standardized certification brought premium prices, and in some cases government subsidies attracted many farmers converting into organic farming. In the developing world, many farmers started organic farming according to traditional methods but are not certified. In other cases, farmers in the developing world have converted out of necessity. As a proportion of total global agricultural output, organic output remains small, but it has been growing rapidly in many countries, notably in Europe.

World Scenario

Across the globe organic farming industry is growing at a substantial pace. The demand of organic food products is especially high in North America and European countries due to increasing health awareness and concerns regarding environmental pollution and conservation. Presently, the global organic food market is estimated around USD 40 billion and growing at a rate of 13%. Organic food is outpacing conventional food industry in terms of growth across the world. Presently, Europe, Japan and US are the largest market of organic food products. European organic food market was estimated around USD 12 billion in 2005 and is growing at a rate of 15-20% annually. While US market alone is estimated to generate sales worth USD 32 billion in 2009. Global Market for Organic Food & beverages is estimated to increase to $ 104.5 billion by 2015 (Markets and Markets Estimate in 2011). In total there are more than 31 million hectares of certified organic agricultural land in the world. The countries with the largest areas of organic land are: Australia, China, Argentina, Italy, USA and Brazil. Some countries such as Liechtenstein, Austria, Switzerland and Italy have reached close to 10% of their agricultural land organic. Amongst African countries Tunisia and Uganda are having more than 1% of their agricultural land under certified organic production. Organic farming has been embraced on a wide scale across various Latin American countries, Asia and Africa. Already China, Turkey and Brazil have become the key suppliers of organic ingredients such as nuts, beans and seeds. While India, Ethiopia and Paraguay have emerged as key supplier of herbs and spices.

Other Asian and African countries have also emerged as key supplier of fruits and other horticulture products which stand mentioned below:

Countries	Percentage of land area under organic management (%)
Liechtenstein	26.40
Austria	13.53
Switzerland	11.33
Finland	7.31
Sweden	6.80
Italy	6.22
Czech Rep	6.09
Denmark	5.76
Portugal	5.42
Estonia	5.17

The ten countries with highest per cent of Organic Area – 2006

Need of Organic Farming

With the increase in population our compulsion would be not only to stabilize agricultural production but to increase it further in sustainable manner. The scientists have realized that the 'Green Revolution with high input use has reached a plateau and is now sustained with diminishing return of falling dividends. Thus, a natural balance needs to be maintained at all cost for existence of life and property. The obvious choice for that would be more relevant in the present era, when these agrochemicals which are produced from fossil fuel and are not renewable and are diminishing in availability. It may also cost heavily on our foreign exchange in future. Hence, organic farming is the need of the day. (Gupta,2019)

Concept of Organic Farming

For many, but especially for policy-makers in developing countries, organic agriculture is a set of strict rules and complicated practices that allow marketing of certified food products. Although traditional farming that are organic by default, that is, this does not use synthetic agriculture inputs nor applies soil building practices and no synthetic inputs qualifies as "Organic". Organic farming is basically one of the farming systems practised with an objective to cultivating and raising various crops without interrupting the composition of soil and thus keeps the soil alive. The process mainly uses the organic wastes such as crop, animal and farm wastes, aquatic wastes and other biological manures and microbes called the biofertilizers which are very friendly to the soil and enhance the fertility of the soil instead of degrading them as the inorganic fertilizers do. The methods and manures used in organic farming are totally eco-friendly and environmentally friendly in nature and hence highly recommended for present time as a lot of damage has already been done to the soil across the world due to excessive use of inorganic fertilizers. With the introduction of inorganic fertilizers, herbicides and pesticides during the industrial revolution to meet the increasing demand of food in the early and mid-90s, use of synthetic fertilizers became quite common across the world leading to the gradual degradation of soil fertility. Though, organic farming is not a new concept of agricultural and has been followed in India since the prehistoric times and is one of the most effective methods of sustainable agriculture of all times. Use of several approaches including inter- cropping, mulching, crop rotation, integration of crops and live stocks etc. in organic farming makes it even better and useful for cultivation of various crops without harming the fertility of the soil. Due to excessive use of synthetic fertilizers the area of fertile land is reducing day by day which is matter of prime concern and to overcome this problem it is very essential to promote organic farming across the globe to retain the fertility of the soil. Organic farming promotes the use of crop rotations and cover crops, and encourages balanced host/predator relationships. Organic residues and nutrients produced on the farm are recycled back to the soil. Cover crops and composted manure are used to maintain soil organic matter and fertility. Preventative insect and disease control methods are practised, including crop rotation, improved genetics and resistant varieties. Integrated pest and weed management, and soil conservation systems are valuable tools on an organic farm. Organically approved pesticides include "natural" or other pest management products included in the Permitted Substances List (PSL) of the organic standards. The permitted substances list identifies substances permitted for use as a pesticide in organic agriculture. All grains, forages and protein supplements fed to livestock must be organically grown. The organic standards generally prohibit products of genetic engineering and animal cloning, synthetic pesticides, synthetic fertilizers, sewage sludge, synthetic drugs, synthetic food processing aids and ingredients, and ionizing radiation. Prohibited products and practices must not be used on certified organic farms for at least three years prior to harvest of the certified organic products. Livestock must be raised organically and fed 100 per cent organic feed ingredients. Organic farming presents many challenges. Some crops are more challenging than others to grow organically; however, nearly every commodity can be produced organically.

Earlier, the concept of organic farming had a very limited and traditional approach but the modern concept of organic farming has come up with expanding horizons and many new concepts. The modern approach of organic farming tends to cover all the related aspects of agriculture that includes soil management, water management, crop production, plant protection and harvesting techniques. This is the reason organic farming is gaining its popularity across the globe very fast. Thus, it is very necessary to adopt the method of organic farming in a large scale to maintain the ecological balance and diversity which is very essential. The land for production is limited and need to be conserved in order to utilize it for a longer period of time. And if the use of inorganic fertilizers will not be checked the amount of fertile land will reduce in a significant rate that can lead to a serious problem of shortage of food for all. Organic farming has come up with a new hope to sustain agriculture in an ecofriendly manner which also provides us with good quality of food without any harmful chemical constituents in it.

Defining of Organic Farming

Organic farming is a method of crop and livestock production that involves much more than choosing not to use pesticides, fertilizers, genetically modified organisms, antibiotics and growth hormones. Organic production is a holistic system designed to optimize the productivity and fitness of diverse communities within the agro-ecosystem, including soil organisms, plants, livestock and people.

As per the definition of the United States Department of Agriculture (USDA) study team on organic farming "Organic Farming is a system which avoids or largely excludes the use of synthetic inputs (such as fertilizers, pesticides, hormones, feed additives etc) and to the maximum extent feasible rely upon crop rotations, crop residues, animal manures, off-farm organic waste, mineral grade rock additives and biological system of nutrient mobilization and plant protection".

FAO suggested that "Organic agriculture is a unique production management system which promotes and enhances agro-ecosystem health, including biodiversity, biological cycles and soil biological activity, and this is accomplished by using on-farm agronomic, biological and mechanical methods in exclusion of all synthetic off farm inputs".

The Key Characteristics of Organic Farming Include

1. Protecting the long-term fertility of soils by maintaining organic matter levels, encouraging soil biological activity, and careful mechanical intervention;
2. Providing crop nutrients indirectly using relatively insoluble nutrient sources which are made available to the plant by the action of soil micro-organisms;

3. Nitrogen self-sufficiency through the use of legumes and biological nitrogen fixation, as well as effective recycling of organic materials including crop residues and livestock manures;
4. Weed, disease and pest control relying primarily on crop rotations, natural predators, diversity, organic manuring, resistant varieties and limited (preferably minimal) thermal, biological and chemical intervention;
5. The extensive management of livestock, paying full regard to their evolutionary adaptations, behavioural needs and animal welfare issues with respect to nutrition, housing, health, breeding and rearing;
6. Careful attention to the impact of the farming system on the wider environment and the conservation of wildlife and natural habitats.

Basic Steps of Organic Farming

Organic farming approach involves following five principles:

1. Conversion of land from conventional management to organic management.
2. Management of the entire surrounding system to ensure biodiversity and sustainability of the system.
3. Crop production with the use of alternative sources of nutrients such as crop rotation, residue management, organic manures and biological inputs.
4. Management of weeds and pests by better management practices, physical and cultural means and by biological control system.
5. Maintenance of livestock in tandem with organic concept and make them an integral part of the entire system.

Why Farm Organically?

The main reasons, farmers state for wanting to farm organically, are their concerns for the environment and about working with agricultural chemicals in conventional farming systems. There is also an issue with the amount of energy used in agriculture, since many farm chemicals require energy intensive manufacturing processes that rely heavily on fossil fuels. Organic farmers find their method of farming to be profitable and personally rewarding.

Why to Buy Organic Foods?

Consumers purchase organic foods for many different reasons. Many want to buy food products that are free of chemical pesticides or grown without conventional fertilizers. Some simply like to try new and different products. Product taste, concerns for the environment and the desire to avoid foods from genetically engineered organisms are among the many other reasons some consumers prefer to buy organic food products. In 2007, it was estimated that over 60 per cent of consumers bought some organic products. Approximately five per cent of consumers are considered to be core organic consumers who buy up to 50 per cent of all organic food.

Successful Organic Farming

In organic production, farmers choose not to use some of the convenient chemical tools available to other farmers. Design and management of the production system are critical to the success of the farm. Select enterprises that complement each other and choose crop rotation and tillage practices to avoid or reduce crop problems. Yields of each organic crop vary, depending on the success of the manager. During the transition from conventional to organic, production of yields is lower than conventional levels, but after a three-to-five-year transition period the organic yields typically increase. Cereal and forage crops can be grown organically relatively easily to due to relatively low pest pressures and nutrient requirements. Soybeans also perform well but weeds can be a challenge. Corn is being grown more frequently on organic farms but careful management of weed control and fertility is needed. Meeting nitrogen requirements is particularly challenging. Corn can be successfully grown after forage legumes or if manure has been applied. Markets for organic feed grains have been strong in recent years. The adoption of genetically engineered (GMO) corn and canola varieties on conventional farms has created the issue of buffer zones or isolation distance for organic corn and canola crops. Farmers producing corn and canola organically are required to manage the risks of GMO contamination in order to produce a "GMO-free" product. The main strategy to manage this risk is through appropriate buffer distances between organic and genetically engineered crops. Cross-pollinated crops such as corn and canola require much greater isolation distance than self-pollinated crops such as soybeans or cereals. Fruit and vegetable crops present greater challenges depending on the crop. Some managers have been very successful, while other farms with the same crop have had significant problems. Certain insect or disease pests are more serious in some regions than in others. Some pest problems are difficult to manage with organic methods. This is less of an issue as more organically approved biopesticides become available. Marketable yields of organic horticultural crops are usually below non-organic crop yields. The yield reduction varies by crop and farm. Some organic producers have added value to their products with on-farm processing.

An example is to make jams, jellies, juice, etc. using products that do not meet fresh market standards.

Livestock products can also be produced organically. In recent years, organic dairy products have become popular. There is an expanding market for organic meat products. Animals must be fed only organic feeds (except under exceptional circumstances). Feed must not contain mammalian, avian or fish by-products. All genetically engineered organisms and substances are prohibited. Antibiotics, growth hormones and insecticides are generally prohibited. If an animal becomes ill and antibiotics are necessary for recovery, they should be administered. The animal must then be segregated from the organic livestock herd and cannot be sold for organic meat products. Vaccinations are permitted when diseases cannot be controlled by other means. Artificial insemination is permitted. Always check with your certification body to determine if a product or technique is allowed in the Permitted Substances List and the organic standards. Organic production must also respect all other federal, provincial and municipal regulations. Organic produce can usually qualify for higher prices than non-organic products. These premiums vary with the crop and may depend on whether you are dealing with a processor, wholesaler, retailer or directly with the consumer. Prices and premiums are negotiated between buyer and seller and will fluctuate with local and global supply and demand. Higher prices offset the higher production costs (per unit of production) of management, labour, and for lower farm yields. These differences vary with commodity. Some experienced field crop producers, particularly of cereals and forages, report very little change in yield while in some horticultural crops such as tree and fruits, significant differences in marketable yield have been observed. There may also be higher marketing costs to develop markets where there is less infrastructure than for conventional commodities. Currently, demand is greater than supply for most organic products.

Advantages of Organic Farming

- Organic farming is more cost effective. It reduces the production cost by about 25-30%, because it does not involve the use of synthetic fertilizers and pesticides.
- It retains 40% more topsoil, thus increasing the crop yield up to five-fold within five years.
- Organic farming is more profitable because it reduces water use, nutrient-contamination by pesticides and reduced soil erosion.

- It also enables the farmers to use the soil for a longer period of time to grow crops maintained for a long time.
- Cattle grazing on organic farmlands have been found to be less prone to diseases and they yield more health milk.
- Products or foodstuffs produced from organic farming do not containany sort of artificial flavours or preservatives.
- Due to the absence of synthetic fertilizers and pesticides, the original nutritional content of food is preserved.
- Organic farming also helps reduce the occurrence of many ailments, and speeds the recovery by boosting the immune system.

Disadvantages of Organic Farming

- Organic farming results in smaller yields, and is more labour intensive, and time-consuming.
- Organic fertilizers tend to release slowly, and hence may need several applications before the desired results can be brought about.
- Farming the organic way requires deep skill and extensive knowledge.

Indian Scenario

Organic farming system in India is not new and is being followed from ancient time. It is a method of farming system which primarily aimed at cultivating the land and raising crops in such a way, as to keep the soil alive and in good health by use of organic wastes (crop, animal and farm wastes, aquatic wastes) and other biological materials along with beneficial microbes (biofertilizers) to release nutrients to crops for increased sustainable production in an ecofriendly pollution free environment. Indian farmers were basically organic farmers before the advent of inorganic fertilizers and chemical pesticides. Overtime the use of these synthetic inputs has come to the level of causing a concern to the environment and human health. Consequently, it is felt necessary to advocate the use of the age-old practice of organic farming not only to ensure uncontaminated food production but also to sustain the agriculture by keeping the land in a healthy condition. In the recent past, this has become a major concern where the consumers started demanding produce grown organically by not using chemicals. To make organic farming successful, it is essential that eco-friendly technologies, which can maintain or increase the agricultural productivity, have to be developed and made available to the farmers.

More than 60% of India's arable land is under traditional agriculture, where no synthetic inputs are being used. Although, the products grown under such systems have so far not been defined as organic products but by all mean they are genuine organic products. In view of their wide availability there is an urgent need to ensure premium prices for the produce grown in these regions. Unfortunately, these farmers are so involved in their struggle for survival that they have no time to figure out what is organic and what is not. These organic products are sold to the middleman and are being marketed along with other chemically grown products. It is the lack of awareness among the consumers in our country that sometimes the chemically grown products which look healthy and attractive, in spite of having alarmingly high level of pesticide residue fetches higher prices than the poorly looking organic products. There is neither subsidy for organic cultivators nor incentives to practise organic cultivation. Area under organic farming has grown many-fold in six years to 2009-10 in India on the back of thrust given to the chemical-free mode of cultivation. From 42.000 hectares under organic certification in 2003-04, more than 4.4 million hectares area was under organic certification in the country as on March 2010, according to an official statement. For quality assurance, India has internationally acclaimed certification process in place for export, import and domestic markets. During 2008-09, India produced about 18.78 lakh tonnes of certified organic products. Of this, nearly 54.000 tonne food items worth Rs 591 crore were exported. Currently, India exports about 86 products worth over 100 million dollars to the world certified organic market registering a growth of over 30 per cent. Certified organic products including all varieties of food products namely basmati rice, pulses, honey, tea, spices, coffee, oil seeds, fruits, processed food, cereals, herbal medicines and their value added products are produced in India. Apart from edible sector, organic cotton fibre, garments, cosmetics, functional food products and body care products are also produced. Banana, pomegranates, pineapple, grapes, ginger, large cardamom, sweet fennel, peanut, onion, sugar/jaggery are other commodities which will emerge as significant organic commodities produced in India in the next two to three years. "In 2007-08, global organic cotton production increased by 152 per cent (145,872 metric tons or 668,581 bales.) "The area under cotton cultivation increased to 161,000 hectares in top organic cotton-producing countries like Syria, India, Turkey and China." Although, organic farming is picking up pace in India but the sector has been jostling with lack of awareness, knowledge and confidence about organic farming, food products among both farmers and consumers.

The Ministry of Agriculture is promoting organic farming in the country under National Project on Organic Farming, National Horticulture Mission,

Technology Mission for North East and Rashtriya Krishi Vikas Yojana. National Project on Organic Farming is being implemented since October 2004 through a National Centre of Organic Farming at Ghaziabad and six regional centres located at Bangalore. Bhubaneswar, Hissar, Imphal, Jabalpur., and Nagpur. The project supports organic input production infrastructure, technical capacity building of stake holders, human resource development through training, statutory quality control of organic inputs, technology development and dissemination, market development and awareness.

Under the National Horticulture Mission and Technology Mission for North East, assistance is provided at rate of 50% of cost subject to a maximum of Rs 10,000 per hectare (up to 4 hectares per beneficiary) for organic horticulture cultivation. Assistance is also provided for setting up vermicompost units at the rate of 50% of cost up to Rs 30,000 per beneficiary. Assistance of Rs 5 lakh is provided to a group of farmers covering an area of 50 hectares for organic farming certification.

Under the Rashtriya Krishi Vikas Yojana, states of India are being assisted for area expansion of organic food crops, capacity building of farmers and organic input production. Organic farming is much native to India but it was almost forgotten after Green Revolution(Gupta, 2019)

World record of Organic Potato Production in Bihar

Mr. Nitish Kumar Chief minister of Bihar proudly displayed the potatoes raised by a farmer in his home district of Nalanda who has earned the honour of setting a new world record of potato production using only organic farming. The farmer has managed to raise 72.9 tonnes of potatoes per hectare beating the previous world record of 45 tonnes per hectare established earlier by the Netherlands.

"This is a very proud moment for all of us that further bolsters our resolve of a second green revolution in the state. Nitish has proved that with proper knowledge and technique, Bihar could attain this goal which is not so distant in future." the Chief Minister said.

J&K Scenario

The farmers of Jammu and Kashmir Union Territory as well as those of Ladakh Union Territory can tap the opportunity of growing demand of different organic products. It is point to mention that many farmers of the Jammu and Kashmir have already adopted new organic agricultural technologies but still Jammu and Kashmir is having low productivity of almost all the crops as has been recently reported by Lal (2022). According to him, there is great need to develop systematic approach and right planning for the development in organic farming in Jammu and Kashmir. It is because demands for organic crops is increasing tremendously. Thus "Organic Farming, can lead prosperity among the farmers of Jammu and Kashmir. However proper strategy is required to promote organic farming in J&K. An integrated approach from government to encourage this kind of organic farming is utmost important.

Promotion of Organic Farming in Jammu Region

Farmers in J&K have been doing farming without the use of chemicals on approximately 32,000 hectares of land across the state since long. But now with the growing demand for organic produce around the globe, the state is keen to take up the practice on a much larger scale. The Sher-e-Kashmir University of Agriculture Sciences and Technology (SKUAST) has enlisted more than 200 farmers who already have organic farms. While these farmers want organic certification for their produce, the varsity will help them to increase the organic yield. Vice-Chancellor, SKUAST, Dr Tej Pratap Singh says he sees Kashmir saying no to chemical way of farming in the near future "Kashmir is a natural place to go organic, the state has a vast potential in that respect. " We already have large swathes of land in the state where farmers are growing walnuts and herbs organically. On 32,000 hectares of land, mostly in the countryside, farmers do not use chemicals at all. Many of them export walnut, saffron and almond and fetch handsome amounts. While the walnut is organic, farmers can even grow saffron organically. "But the need of the hour, Singh said, was to take up the practice on a much larger scale. Kashmir produces around 87,000 tonnes of walnut kernels annually from its 40 lakh walnut trees across the state. While walnuts are organic as no pesticides and fertilizers are used to grow them and saffron, which is cultivated on more than 3,000 hectares, is only ten per cent organic. SKUAST has also recently launched a full-fledged organic agriculture programme.

JK Govt. Signs MoUs (Memorandum of Understanding) for Promotion of Organic Farming in Jammu

To promote organic farming in Jammu province, Jammu and Kashmir government signed a Memorandum of Understanding with three companies to help cultivators to switch over to this mode of farming. The agreement was signed by Minister for Agriculture, Ghulam Hassan Mir with M R Morarka GDC Rural Research Foundation (Jaipur). International Panaacea Ltd (New Delhi) and Sarveshwar Organic Foods (Jammu). Under the MoUs at the first instance the companies would help to introduce the organic mode of farming over a sown area of 800 hectares land across the region, Mir said.

The salient features of the MoU include promoting group organic farming by arranging conversion of conventional fields/farms to organic, besides ensuring necessary technology transfer to farmers, certification by agencies approved by Agriculture and Processed Food Export Development Authority (APEDA) and assistance to the farmers in marketing of the organic produce. The Service Providers shall discharge their responsibilities by motivating farmer groups/ clusters by enforcing internal control system, besides acting as technology messenger and advice farmers on various issues. The minimum period of conversation into organic farming shall be three years.

Jammu and Kashmir Union Territory being hilly state, possesses a lot of potential for organic farming. There is no doubt several steps have been taken by the State Department of Agriculture under Jammu and Kashmir. Government in order to highlight the importance of organic agriculture in the growth of the state yet there is need to make rigorous efforts to promote organic farming. It is more so in case of Shivalik's of Jammu (Kandi Belt) and Karewas of Kashmir. It is because soil health of these areas has very much deteriorated due to the participants both farmers (men and women) and orchardists are required to provide suitable fuel wood trees, fruit trees and different grass species depending up on the areas for planting. The main aim of this programme is to provide modalities of protection of both animals and human beings and also from the ravages of fast flowing stream (Nallah′s) during the rainy season, when soil along with many plants get washed away leaving only waste /degraded land having lot of stones and pebbles.

This model is accepted by the authorities, can be utilized for treating the hundreds of such Nallahas which exist in many areas of Jammu region of Jammu and Kashmir Union Territory, Union Territory of Ladakh, Shivalik's of Jammu, Punjab Haryana, Himachal Pradesh and Uttarakhand.

The farm women of the areas of the above said states can be addressed and motivated fully on afforestation programme of Social Forestry/Agroforestry and in the process also make pocket money for themselves by supplying / selling plants from the kitchen nurseries that they can start themselves. Medicines and vaccinations for diseases which once thought was incurable are now no more a cause of worry, which has been a blessing for human kind.

Industrial Research and Development

Green Revolution (Hari Kranti) has been one of the greatest achievements since the second World War. The phenomenal increase of research based agricultural, productivity applied in "Green Revolution" has fed millions of people and served as the basis of economic transformation in many developing countries including Indian subcontinent. There is no doubt that this "Green Revolution had enabled the country to shun predicting death offarmersand occurence famine. However, Both the government and private sectors, employ biotechnologists. They conduct or resort research and development work, for increasing productivity, improving energy production and conservation, minimizing pollution and industrial waste etc. Biotechnologists also find opportunities at places involving activities like chemical processes, genetic engineering, textile development, cosmetic development etc.

Agriculture and Animal Husbandry

Indian economy is very much, dependent up on agriculture and biotechnologists have made major advancements in this area. Over the years, the agricultural output has been improving outing to the improvements in the quality of seeds, insecticides, fertilizers etc. This has resulted in the deterioration of the physical, chemical, and biological health of the rice-wheat growing soils. Hence, currently, there is a growing concern about the sustain ability of the rice-wheat cropping system as the growth rates of rice and wheat yields have either become stagnant or have declined in a number of states such as Punjab, Haryana, Western, Uttar Pradesh, Madhya Pradesh, Bihar, Himachal Pradesh, Madhya Pradesh, Bihar, Himachal Pradesh, Uttarakhand and Jammu and Kashmir (Mahajan et al 2008, Mahajan and Gupta, 2009).

During Green Revolution high yielding varieties of rice and wheat were grown which responded not only to more doses of fertilizers but also those of pesticides and requirement of large quantities of water. They have changed the environmental conditions. Intensive use of pesticides in rice-wheat producing states, Punjab, Haryana and western Uttar Pradesh which reaped the main benefits of Green Revolution,the fact that the farmers are liberated from their dependence on fertilizers and pesticides in their produce.

- All said and done, rely on agroforestry practice. Do not pursue agriculture and silviculture but should be done jointly on farms. Remember that agriculture is meant for the plains while silviculture is for mountains.
- Largescale plantation of quick growing leguminous trees on roadsides, wastelands and field bunds should be encouraged.
- There should be a ban to use cow dung as fuel as has been done in China.
- Biofertilizer technology needs to be refined for easy adaptability by the farmers.

- Developing simple techniques to handle bulky organic manures so that their processing, use and application become as simple as chemical fertilizers.
- A belief that organic food is tastier. Research in this line should be carried out in details as has been done by the soil scientists of the UK soil Association.

References

Acharya, C.L. and Kapur, O.C. 2001. Using Organic Wastes as Compost and Mulch potato in low water retaining hill soils of North West India. Glimpses of the published work in Indian Journal Agricultural Sciences (Personnel Communication).

Acharya, C.L., Ghosh, P.K and Subba Rao, A. 2001. Indigenous Nutrient Practices-Wisdom alive in India. Indian Institute of Soil Science, Bhopal, M.P. (India).

Acharya, CL, Sood, M.C and Sharma, R.C. 2001.Practices in Himachal Pradesh. In Indigenous Nutrient Management Practices Wisdom alive in India (Ed. Acharya, C.L., Ghosh, P.K. and Subba Rao A) Indian Institute of Soil Science, Bhopal, M. (India).

Anonymous, 2005. Indian Organic Products Certifying bodies to get EC status. AgriNews-Farmers Forum. 5(11): 28

Bala Subramanian, C. 2000. Friendly alternatives to chemical fertilizers and food grain production. Himsa Research Foundation Ahinsa Sthal Mehrauli, New Delhi, India.

Bhardwaj, K.K.R; Kalyana Sundaram, N.K. and Khan, RH. 1998. Management of organic materials for field and plantation crops. In Bulletin. Indian Soc. Soil Science. 19:122-134.

Burcher, Sam. 2007. FAO promotes of organic Agriculture. The Kashmir times 64(320):10 [November, 24 Saturday].

Chhonkar, P.K. 2001. The state of Indian soil science and challenges to be faced during twenty first century. 1. Indian Soc. Soil Sci. 49: 532-536.

Chhonkar, P.K. 2003. Organic farming: Science or belief. 1. Indian Soc. Soil Sci. 51: 365-377.

Dahatonde, S. and Khambalkar, V.P. 2007. Use of Organics in Indian Agriculture. Agrobios. 5(12): 38-39.

Dutt, S. 2002. Organic Farming for Sustainable Development. The Daily Excelsior. 38 (251):7. Gupta, R.D.2019.Organic farming need of the day. The State Times.24(4);6

Lal Banarsi. 2022. Promoting organic farming in J&K. The state Times 27 (90): 6

Mahajan, A. and Gupta, R.D. 2009. Integrated nutrient management (INM) in a sustainable rice-wheat cropping system. Springer Science & Business Media (www.springer.com)

Mahajan, A., Gupta, R.D. and Sharma, R. 2008a. Biofertilizers-A way to sustainable agriculture. Agrobios Newsletter. 6(9): 36-37.

4

Organic Farming and Horticulture

Organic farming in India started receiving focused attention from the year 2004-2005 when National Project on Organic farming (NPOF) was launched during 2004, the area under organic farming was 42,000 hectares (Srimathi and Jency, 2014). By March, 2010, it was reached to 1.08 million hectares. In addition to these, 3.40 million hectares was wild forest harvest collection area. Thus, total area under organic certification process by March,2010 was 4.48 million hectares which was increased by 25 folds during last 6 years. In cultivated organic land 7.56 lakh hectares was certified while 3.2 lakh hectares was under conversion. In India, States such as Orissa, Jharkhand, Uttarakhand (Uttaranchal), Himachal Pradesh, Jammu and Kashmir Rajasthan, Gujarat, Madhya Pradesh and Chhattisgarh have been categorized as low or negligible fertilizers consuming states of the country.

The states doing well in organic farming mainly comprise of Madhya Pradesh (4.40 lakh hectares), Maharashtra (1.50 lakh hectares) and Orissa (95,000 hectares) which have the largest area under organic farming. Among the crops, cotton is the single largest crop accounting for nearly 40 per cent of the total area under organic farming followed by rice, pulses, oilseeds and spices. It is remarkable to note that India is the largest producer of organic cotton in the world, and accounts for about 50 per cent share of total world organic cotton production.

Organic farming is being promoted under various scheme viz, National Project on Organic Farming (NPOF), National Horticulture Mission (NHM) and Rashtriya Krishi Vikas Yojana (RKVY).

Why to Buy Organic Food?

Consumers want to buy food products which are almost free from chemical pesticides and fertilizers. It is because these food products are grown without the use of chemical fertilizers and pesticides. Some people simply like to try new and different products. Product taste concerns for the environment and desire to avoid foods from genetically engineered organisms are among the many other reasons. Organically produced products are in very good demand especially in foreign countries. APEDA, Manufacture of commerce and industries, Government of India is granting trade mark of the "Indian Organic Agriculture" on the basis of compliance with national standards of organic production (Sachan et al. 2007).

What is Meant by Certified Organic Foods ?

"Certified Organic Food" is a term which is given to food products produced in accordance with organic standards as is certified by one of the certifying bodies. There are a number of certification bodies which are operating in India. A grower wishing to be certified for its organic produce, has to apply to a certification body requesting an independent inspection of his farm to verify that farm meets all the necessary organic standards framed by the inspecting bodies. Farmers, processors and traders are each required to maintain a document trial for audit purposes. Products from certified levelled and promoted as, "Certified Organic". An organic certification is necessary to prove that one's produce is organic in nature.

National Programme on Organic Production (NPOP)

NPOP can be defined as regulatory mechanism and is regulated under two different acts for export and domestic markets. Accreditation of certification and Inspection Agencies is being granted by a common National Accreditation Body (NAB) and 18 accredited certification agencies are looking after the requirement of certification process. Out of these 4 agencies are under public sector while 14 are under private management, organic farmers and operators in India.

The Transition Period

The first few years of organic agriculture production are highly tedious / difficult. Organic farming standards require that lands to be cultivated for organic farming must be well managed using the organic practices i.e., either use of well decomposed organic manure or compost, vermicompost and green manuring at least for 3 years prior to harvesting of the first certified organic crop. This is known as transition period when both the soil and manager adjust the new system of farming i.e., "Organic Agriculture" or "Organic Farming. Insect, pest and growth of weeds also require to be adjusted during this time. Prepare a plan for adoption of organic farming with full swing. Try to adopt 10- 20 per cent organic farming programme during the first year. Pick up one of the best cultivated fields to start with the aforesaid organic farming. And then expand organic acreage as knowledge and confidence are gained.

It may take five to ten years to become fully or totally organic producer, but a long-term approach is often successful than that of a rapid conversion.

Successful Organic Farming

In organic farming production, the yields of each of organic crops vary depending up on the success of the manager (Farmer). During the transition from conventional to organic production, yields of the crops are lower than conventional levels. However, after a three-to-five-year transition period, the organic yields increase typically. Organic cultivation of cereal and forage crops, is easy due to relatively low pest pressures and plant nutrient requirements. Soybean crop also performs but the occurrence of weeds can be a challenge. Corn is being grown frequently on organic farms but careful management of weed control and maintenance of soil fertility is very much needed. It is point to mention that corn can be successfully grown after for age legumes or else organic manure has to be applied.

Fruit and vegetable crops indicate greater challenges. It is attributed to the fact that certain insect or disease pests are more serious in some regions than others. Marketable yields of organic horticultural crops are generally below the non-organic crop yields. However, some organic horticultural producers have now started to use organic pesticides made up from Neem (*Azadirachta indica*) tree.

References

Sachan, V.K, Manral, H.S, Kumar, R and Singh, D. P. (2007). Organic farming - An Eco-friendly way of sustainable production. Indian Farmers Digest, 40(4):27-29

Srimathi,S and Jency, J.P. (2014). Organic Farming - Natural way of Farming. Agrobios Newsletter. 13(5):62-64

5

Organic Vegetable Farming

India is the second largest producer of vegetables after China. The diverse agro climatic conditions of India offer an immense scope for production of organic vegetables throughout the year. The increase in area, production and productivity associated with nutrition and health security awareness demands lot of promotion of organic vegetables cultivation in the region since these products sell at the premium prices in the domestic and overseas market (Sandhu and Khera, 2008). It is because the major contribution of vegetables to human health is the large amount of Vitamin C and A as well as folic acid and good number of dietary fibres and minerals (Sanwal, 2008). Apart from this, the phytochemicals in vegetables protect the human body from cancers and cardiovascular diseases. Under organic mode of cultivation one can get an average yield of 25-30 tonnes of good quality potato vegetable crop per hectare (Sachan et al. 2008). It is worthwhile to mention that the potato is a key world food vegetable crop which is produced in 130 nations. It is grown from sea level to 3900 m elevation and its edible dry matter accounts for a higher volume of the food consumed in the world than fish and meat combined together (Sachan et al. 2008). More potatoes are consumed by the world's population than any other vegetables.

The potato's incredible variety as well as its high-Andean origin, makes fascinating crop to grow organically. It is believed that it has much potential for sustainable, non-chemical farming fits in well with many crop rotations, and does very well with natural or organic manures. Crop response to application of biofertilizers has also been observed (Sachan et al. 2008). Application of well rotten farm yard manure (FYM) @ 20 tonnes ha^{-1} or vermicompost @ 10 tonnes ha^{-1} can meet nutrient requirement of the potato crop. The biofertilizers namely *Azotobacterin* and *Phosphate solubilizing bacterial cells* can also help in reducing the nitrogen and phosphorus requirement of the crop.

Use of Mulching in Vegetables Production

Mulching is a practice which is often used by organic vegetable growers. Traditionally, mulching entails the spreading of large amounts of organic materials straw, old hay, wood chips etc. over otherwise bare soil, between and among crop plants / vegetables. Organic mulches regulate soil moisture and temperature, supresses weeds and provides organic matter to the soil. Organic matter on decomposition provides humus to the soil. In cultivated soils as much as nine tenth of the soil organic matter may consist of humus which in intimate mixture with inorganic colloids is called clay humus complex. Organic matter helps to buffer soils against rapid chemical changes in pH due to addition of calcium and magnesium. Humus provides a store house for available nutrients. Organic matter also serves as a source of energy for the growth of soil microorganic. Organic matter ameliorates the physical condition of the soils, enhances cation exchange capacity and water holding capacity of the soils.

In vegetable crop rotations, nitrogen fixation and carry over is also very important. (Tiwari, 2007). An eight-year rotation mentioned by Tiwari (2007) is as under:

- Potatoes follow sweet corn because research has shown corn to one of the preceding crops that most benefit the yield of potatoes.
- Sweet corn follows the cabbage family because in contrast to many other crops corn shows no yield decline when following a crop of brassicas. Secondly, the cabbage family can be under sown to a leguminous green manure which when turned under following spring provides the most ideal growing conditions for a sweet corn.
- The cabbage follows peas because the pea crop is finished and the ground is cleared (early) allowing a vigorous green manure crop to be established.
- Peas follow tomatoes because they need an early seed bed and tomatoes can be under sown non winter hardy green manure crop that provides soil protection over winter with no decomposition and regrowth problem in the spring.
- Tomatoes follow beans in the rotation because this places them 4 years away from their close cousin the potato.
- Beans follow root crops because they are not known to be subject to the detrimental effect that certain root crops such as carrots and beets may exert in the following year.
- Root crops follow squash and potato because these two crops are good cleaning crops, as there are fewer weeds to content within the root crops. Squash has been found to be a beneficial preceding crop for roots.

- Squash is grown after growing of potatoes in order to have the two cleaning crops back-to-back prior to the root crops, thus reducing weed problems in the root crops.

The Union Territory of Jammu and Kashmir has decided to establish, "School Nutrition Garden" in all the government schools of Jammu and Kashmir to use fresh organic vegetables grown in such gardens for the preparation of mid day meals for the children as reported by Sharma (2020). This movement is also aimed at involving children with natural organic farming besides cultivation of different kinds of vegetables among the school children. The vegetables grown in these school gardens will be utilized to prepare Midday Meal Scheme. The idea behind the competition is to promote community participation and the children interface with the "Nature". The aforesaid idea was given by the Director Mid-Day Meal (MDM) Scheme of Jammu and Kashmir.

It is point to mention that Union Territory of Jammu and Kashmir has lot of potential for organic farming as a large area in Jammu and Kashmir is already under semi-organic cultivation especially in hilly districts of the J&K due to lack of the availability of chemical fertilizers in these areas and the farmers of these areas hardly apply the chemical fertilizers (Lal, 2022). Many of the farmers of Reasi district in Jammu grow vegetables like turmeric and ginger purely organically. Similarly, a large number of farmers of Gurez of Kashmir grow potatoes totally in an organic way.

In the end it is concluded from the foregoing literature that the farmers / orchardists must grow vegetables in an organic way. It is because of the following reasons

1) Vegetables do not only adorn the table but they also enrich the health of men, women and children. As a matter of fact, vegetables form the most nutritive menu of man and tone up the energy and vigour.

2) Growing and eating of vegetables were appreciated by the ancient people for their tempting acculence, pleasing flavours, high nutritive value and regulatory effects (Singh, 2023).

3) Regular use of vegetables provides or supply many of the most essential health building and protective substances such as carbohydrates, vitamins and minerals, which are generally lacking in other food stuffs.

4) Those who use vegetables less or those who are not in a position to afford them always suffer from minerals and vitamins deficiency diseases.

5) Although vitamins occur in small quantities in vegetables, they produce

profound and specific physiological effects, if eaten properly and regularly.

6) If vegetables are used properly and regularly, they provide us a clear and soft skin and bright eyes.

7) Some of the vitamins found in the vegetables are vitamins A, B, C, D, E and Vitamin G. All of these vitamins are very essential for various physiological activities of the human body like growth, reproduction, strengthening of the bones etc.

8) The deficiency of vitamin A for example causes night blindness, formation of stones in kidney and bladder, dryness, pimpling eruption of skin etc.

9) The deficiency of vitamin C causes unhealthy gums, tooth decay, loss of energy, delay in wound healing, some heart diseases etc.

10) The deficiencyof Vitamin D produces rickets and weakening of bones etc. The deficiency of vitamin E causes sterility and reproductive disorders.

11) At least ten mineral elements are needed for the proper growth and development of human body. Out of these ten elements, calcium, iron and phosphorus are required in large quantities and these are not present in sufficient amount in other food articles except in vegetables.

12) Vegetables especially "Leafy green" such as spinach and broccoli, contain vitamin B6 and folic acid which help to produce serotonin, a brain chemical that boosts one's mood and alertness. Apart from this, vegetables serve as brain tonic.

13) Leafy greens such as kale as well as turnip, spinach dandelion greens and Swiss Chard are all amazing foods that provide iron and lot of vitamin C both are good for strong teeth and hair (Anonymous, 2012)

14) Human eyes require vitamin Ato prevent night blindness which is mostly present in leafy green vegetables (Gupta, 2020). Hence one must take green leafy vegetables. Drumstick leaves fenugreek leaves and radish leaves are some of the examples from this angle. Hence, night blindness sufferers must include aforesaid vegetables in their diet.

Now a days, Indian Agriculture is in deep crises. Such crises consist of both human and ecological nature (Gupta, 2010). Two dimensions of human crises are the suicides of the farmers and growth of hunger and malnutrition. The agrarian crises leading to farmers suicides is an occurrence of debt and this is the result of the rising prices of non-sustainable and inappropriate production inputs and less benefits to the farmers/peasants due to the unfair trade pattern (Gupta *et al.*, 2022). Costly seeds and chemicals (fertilizers and pesticides) drain the peasants, scare capital and leave agroecosystem more fragile and impoverished and the following of the monoculture of the crops i.e. growing of rice and wheat year after year further aggravate the risks of crop failure due to pests and diseases and climate change. It is concluded that the quality of the agriculture produces particularly of vegetable, fruits and flowers improve while plant nutrients are applied organically than the inorganically.

References

Anonymous. 2009. Food for fertility. The State Times.14(98):6

Anonymous. 2012. Brain Food- Eat your way to a sharper mind. Times of India-Times of Jammu. 3(18):4

Gupta. R.D., Gupta, S.K. and Mahajan, A. 2022. Sustainable system of farming for modern agriculture. NIPA Genx Electronics Resources and Solutions Ltd. New Delhi.

Gupta R.D. 2020. Eat Green Leafy Vegetables to Prevent Blindness. The State Times.25(35):6 Gupta R.D. 2010. Organic Agriculture Key to Food Security. The State Times.15(161):6

Lal, Banarsi. 2022. Sustainable Agriculture Through Organic Farming. The State Times 27 (235):6

Sachan, H.K, Vishwakarma, V.K. and Krishna, D. 2008. Successful production of organic potato. Indian Farmer 's, Digest. 41(8):31-33.

Sandhu, Savreet and Khehra, K.S. 2008 Organic Vegetable Farming: Nutrition and Health

Sanwal, S. K. 2008. Underutilized Vegetable and Spice Crops. Agrobios (India) Behind Nasrani Cinema Road, Jodhpur Rajasthan, India.

Sharma, Arteev. 2020. Jammu and Kashmir Government Schools to Grow Organic Nutrition Garden.Tribune News Service,20 Jan.2020

Singh, Arvind Prakash, 2023. Vegetables for all. Personnel Communication on the Address Prof. (Dr). R. D. Gupta 258-C, Sainik Colony, Jammu – 180011, J&K.

Tiwari, Akhilesh 2007, Organic Farming of vegetables. Agrobies Newsletter. 6 (7) 339-41.

6

Organic Medicinal Plants Farming

Medicinal and aromatic plants constitute a precious resource for mankind. Since time immemorial these plants have been put to medicinal use by Hakims, Vaidas, Ayurvedic practioners and of course the common man (Gupta *et al.* 2016). It is because these plants possess a lot of qualities for curing the most common or dreaded diseases without any side effects (Gupta *et al.* 2016). Unfortunately, with the onslaught of human population explosion, industrialization, fertilization and deforestation, the rich medicinal plants wealth has suffered a massive damage. A large number of medicinal plants are now at the verge of extinction which requires their preservation.

India which stands blessed well vast medicinal plants diversity is estimated to have about 30,000 different plant species world over having accepted medicinal value in Ayurveda, Unani, Homeopathy, Allopathy, Tibetan etc. In fact, out of the 18 hot spots in the world with regard to occurrence of medicinal plants two are confined to India. These two hot spots are represented by the Eastern and the North western Himalayas, and the western Ghats with very high degree of endemism. These are indeed the treasure house of about 30,000 or around 45,000 medicinal and aromatic plant species pertaining to medicinal culture (Gupta et al. 2016). Out of these nearly 6000-7500 plant species have been very much exploited for the last 5,000 years traditional system to medicines and are likely to extinct.

Indian medicinal plants/herbs grown naturally or cultivated, have been used in Ayurveda, Unani and Siddha, should always have natural origin and free from any chemicals harmful to the human' s health (Anonymous 2003). Similarly, the medicinal plants must always be grown organically i.e., without the use of any chemical fertilizer and pesticides, weedicide, fungicide, bactericide as well as insecticide and rodenticide.

Thus, nowadays, there must be a distinct revival of global interest in organic botanical, natural herbal medicines, like organic agricultural, horticultural, olericulture products a n d production of herbal medicines organically is the current wave of green consumerism. We should, therefore, definitely consolidate on the goal of better health healing and thereby earning of teeming millions. It is because India has remained and still stands as one of the largest countries in medicinal plants production and their properties (Gupta et al. 2016). Apart from this, there is huge potential of organic substances which still remains untapped to be utilized for production of medicinal plants organically.

The wastes can be converted into organic manure. The use of green manure and biofertilizer has added advantage to the production of medicinal plants production system. Vermicompost and biofertilizers viz.,Rhizobium culture, Azotabacterin, Phosphobacterin and Algal cultures which influence the physical, chemical and biological properties of the soils. They improve the water holding capacity of soils and play an important role in building up of plant nutrients in soils and thereby producing organic herbal medicinal plants products. The organic farming though is hard to practise yet the sincere and cooperative efforts in this direction will be fruitful and our country can achieve enviable position in the world in Ayurveda. It will also assist to increase the Indian Herbal Drug market which is presently around one billion dollars. It is worthwhile to mention that the demand for organic medicinal, products is looming very fast and by adopting the advantage of rising green consumption our country can acquire unviable position in the world.

While growing medicinal plants organically, the growth of weeds can be checked by an integrated approach using combination of cultural and direct control methods like mechanical hoes flushing out germinated weed seeds mulching and solarisation. In organic cultivation of medicinal plants biological control agents can also be used. Herbal extract viz., Neem (*Azadirachta indica*) can be used more successfully under plant protection measures. in disease control Cultivated cultural techniques combined with microbial actively, presence of beneficial root colonizing bacterial, mycorrhizal fungi has remained one of the contributing factor.

Thus, in the changing scenario of green organic food production and consumption in the world our country must always take an advantage and concentrate on green or organic and/ or natural farming.

Growing of Medicinal Plants in the Siwalik Range of Himalayas

The whole of the Siwalik range of Himalayas including Kandi belts of Jammu, J&K State, Punjab, Haryana, Himachal Pradesh and Uttarakhand was once worldwide famous for growing of some medicinal plants as well as bamboo cultivation and fruit trees particularly local mango, phalsa, loquat (Jamwal, 2007). But over exploitation of these natural resources due to burgeoning human and livestock population, has now depleted this wealth very badly. Even a number of villages have degraded and tirelessness prevails all over there. Medicinal trees like Neem, Amla, Harar, Bahera and others were rarely seen about two decades ago. Many natural fruits and vegetables of the forest, which used to flood the market were not available these days. Due to lack of medicinal trees/plants the people of the area, especially of Kandi belt of Jammu now take up allopathic medicines to cure the diseases which have caused a number of side effect. Hence, to get rid of side effects of allopathic medicines, the people of the area have now advocated to boost the production of medicinal plants, especially in the degraded forest lands and badly eroded agricultural lands.

There is distinct revival of global interest in botanical natural products and herbal medicines (Anonymous,2005). In the current wave of green consumerism, we should consolidate on the goal of providing better health healing and caring of teeming millions.

Our country is the source of the ancient literature on the properties of medicinal plants as well as rich biodiversity. Out of the eighteen hot spots in the world two are in India. These are as follows:

1. Eastern Himalayas
2. Western ghats with very high degree of endemism

As a matter of fact, these are the treasure house of around 45,000 plant species. Of these around 6000-75000 plant species have been exploited for the last nearly 5000years traditional system of medicines. Indian medicinal herbs used in Ayurved, Unani and Siddha should have natural origin and free from any residue of chemicals harmful to human health.

Out of the various medicinal plants first grown in the nursery and thereafter in the fields, the following medicinal trees/ plants have proved very well to grow in the area (Jamwal; 2007).

Local name of the tree/plant	**Botanical name**
Amaltas	Cassis fistula
Amla	Phyllanthus emblica
Arjun	Terminalia arjuna
Ashwagandha	Withania somnifera
Bael or billan	Aegle marmelos
Bahera	Terminalia belerica
Barian	Acorus calamus
Bana	Vitex negundo
Vasakar or Braihkar	Adhatoda vasica
Harar	Terminalia chebula

Local name of the tree/plant	**Botanical name**
Kanwar Gandhal	Aloe vera
Kikar	Acacia nilotica
Khair	Acacia catechu
Mulathi	Glycerrhiza glabra
Neem	Azardirachta indica
Pla or Dhak	Butea monosperma
Pulai	Acacia modesta

Long lago many communities of Hindus and other religions used to grow group of plants/trees known as sacred groves. Despite not much interest by the young generations, the sacred groves are still being maintained by some communities of Rajasthan and many others dwelling in north east hills, north west Himalayas vis-a-vis central and south India (Gupta, 2009). The concept of sacred groves dates back to the Prevedic times. Traditionally worshipping of nature has been part and parcel of Hindu's life. Accordingly, the pipal tree (*Ficus religiosa*) and Tulsi plant (*Ocimum sanctum*) are planted in large areas as sacred groves serving as religious spots.

In one of the Shaloks of Atharveda 11, it is stated that banks of the rivers streams, lakes and ponds must be shaded by planting trees like Amra i.e, Mango (*Mangifera indica*),Jambu or Jamun - black berry (*Sygygium cumini*), Borh or Banyan tree (*Ficus bengalensis*) and many others. These trees/plants besides having religious importance possess lot of medicinal values. As for example Jambu or Jaman tree is well known to have lot of power to cure diabetic patients after consuming its fruits or powder of seeds. Similarly, the tree species like Pipal, Borh, Neem and Drek (*Melia azedarach)* have a number of medicinal values, as have been recently reported (Gupta et al 2016). According to Sharma (2014) the main traditional medicinal plants of Jammu region which were tested for their anticancer efficiency and suppressing the proliferation of different cancer cells from different origin are as follows:

1) Pudina (*Mentha aquatica*)
2) Karela (*Momordica charantia*)
3) Karipata (*Murraya koengii*)
4) Tulsi (*Ocimum sanctum*)
5) Adark (*Zingiber officinale*)

Like other trees and plants, aforesaid species of plants can be grown individually in kitchen garden.

References

Anonymous. (2003). Zargar asks saffron producers to adopt modern technologies. The Daily Excelsior. 39 (305):5

Anonymous. 2005.Organic revolution and cultivation of medicinal plants. Farmer's Forum. 5 (6):16

Gupta, R.D. 2009 Saffron – A wonderful medicinal and aromatic plant. Indian Farmers Digest, 42(11):8-11

Gupta, R.D; Gupta, S.K. and Bhardwaj, S.D (2016). Agrotechniques and uses of Medicinal Plants. Associated Publishing Company, A Division, of Astral International Pvt. Ltd ISO 9001: 2008 Central Company 476061/23, Ansari Road, Darya Gang New Delhi-110002

Jamwal, J.2007. Medicinal aromatic plants and value addition. The Kashmir Times. 64(338):15.

Sharma, O.P. (2014). Medicinal Plants of Jammu and Kashmir. Directorate of Social Forestry.

7

Organic Farming and Food Security

In India, Mahatma Gandhi known as father of the Nation, deserves for imparting credit as pioneer for organic agriculture by way of promoting farmyard manure, compost and advocating use of night soil to enhance the soil fertility and thereby, crop productivity. Organic agriculture possesses lot of advantages over conventional agriculture, especially in areas having low rainfall and showing low soil fertility (Yadav, 2010). Although APEDA (Agricultural and Processed Food Products Export Development Authority) as an apex body which is working to promote the export of organic food commodities. However, lack of adequate certification facilities and the trainings organized, markets for organic foods inadequate, political and scientific will power are some of the hindrances for operating organic agriculture in India. As human and natural resources are readily available in our country, so there is good scope for organic agriculture farming including organic animal husbandry by adopting proper organic agriculture policies.

Organic agriculture is a production system that sustains the health of soils and other ecosystems like people, livestock and wildlife. Organic farming consists use of composting, green manuring, crop rotation, diversification of agriculture, biological pest control, use of biofertilizers as well as zero tillage. It avoids the use of chemical fertilizers and pesticides as well as plant growth regulators and food additives. It also excludes or limits the use of antibiotics in live-stock feed as growth promoters.

Organic Agriculture and Food Security

Organic agriculture and food security are debatable especially in developing countries, having certain apprehensions. Organic agriculture is definitely ecofriendly and favourable for bio diversity and health of environment including soil, human and animal health. As organic agriculture revitalizes the biological as well as chemical properties of the soil which as a result sustain yields of different crops grown in the fields. In spite of the fact that the net profit to the farmers may not be declined in organic agriculture. It is attributed to reduction in the cost of production and higher price to be fetched by the organic produce, the apprehensions regarding protection of crops against insects pests and diseases require to be addressed on priority through suitable cost effective alternatives which are discussed below

1) Some of the afore said alternative products: Ingredients and strategies nclude the use of organic green pesticides like neem-based preparations. Neem based preparations are also called "Ecofriendly Botanical Pesticides".

Neem (*Azadirachta indica*) is the most promising source of biopesticides of biological origin. Neem owes its toxic attributes to a large number of bitter compounds called *meliacins, azadirachtins, nimbin, salanins, meliantriol* etc. Among them *Azadirachtin* is the most potent one. Neem seed kernels are the richest source of *meliacins* which contain 0.2-0.3 percent azadirachtin and 30-40 percent oil. Although neem leaves, seed coats and bark also contain *meliacins* and *azadirachtin* but their percentage is very low.

The neem products affect the pests by functioning as feeding and oviposition deterrent, insect growth regulators, chemo sterilant and toxicant. Any pest escaping one effect, may get killed by other effect. The neem has been found effective against more than 200 pests including locusts, grasshoppers, hairy caterpillars, stem borers, mealy bugs. white flies, fruit flies, mosquitoes and various pests of stored products.

2) Other Botanical Pesticides: The other botanical pesticides like pyrethrum rotenone, ryania and nicotine were in vogue prior to the advent of synthetic pesticides. But later on, these botanicals were relegated to the significant position in pest control. Pyrethrum is extracted from dried flowers of chrysanthemum (*Chrysanthemum cinerariefoliím*). It has a rapid knockdown effect on flying insects and is very effective against house hold pests.

3) Adoption of Multiprong Approach: Biopesticides have found of immense utility in organic farming catering to niche market as there is no other way of managing pest problems. Most critical aspect in organic farming is the protection of crop or insects, pests and diseases happened to occur due to viruses, bacteria, fungi etc (Khan, 2010). To accomplish this, a multiprong approach including the use of biopesticides as already stated, herbal and plant-based preparations, pheromones, light traps, animal dung and urine-based products along with the use of pest resistant varieties of crops through well planned plant breeding department by way of utilizing biotechnological tools must be tried and standardized for various crops in different geographical areas of the country.

Mitigating of Climate

Organic farming is helpful in mitigating climate change by way of reducing emissions and increasing carbon sequestration, reduction in the leaching of nutrients from the soil and protecting biodiversity of plants, animals and biota.

Development of Standards for Organic Farming

There is urgent need to develop standards for organic farming or organic agriculture commodities along with their carbon foot print. There is great need for creation of special department for undergoing research on an advanced organic agriculture especially for doing more research on organic livestock strategies. Thus, organic agriculture can be almost carbon neutral

It has emission reduction potential to extent of 20% of global agricultural greenhouse gas emissions besides soil carbon sequestrations up to 40-72% of global agricultural greenhouse gases.

Popularity of Organic Farming

Although organic farming has gained much popularity in both developed and developing countries, yet its operators/controllers are different. While the controllers of organic agriculture in the developing countries include higher income, exports, natural resources conservation, employment generation reduction of greenhouse gas emissions and investment by the private sector. Contrary to this, the developed countries find in organic agriculture economically, ecologically and socially sound option to reduce surplus food commodities. For example, the European Union aims at sustainable agriculture and rural development through organic agriculture. Market demand and social pressure are responsible for continuous enhanced option of organic agriculture both in developed and developing countries. Not only this, there is an increase in the demand for organic food, consumer demand for natural cotton textile is also increasing over the synthetic one.

About 90 developing countries export certified organic food products to markets in Europe, Japan and North America. There is an urgent need to integrate agricultural policies with the organic agriculture policies production along with strengthening of research to achieve higher production through organic agriculture. Current agriculture support policies are not so advantageous to organic agriculture. Creation of certification authority by the government and arrangement for subsidies to cover at least 30% of production inputs and 70% of certification costs for a period of at least 5 years along with easily available farmers need consideration by the planners and policy makers. Now there is the growing demand for organically farmer produce both for the consumer and the producers. It is attributed to more nutritious the organically grown foods as compared to conventionally grown foods (Sharma, 2022).

References

Chander, S and Munshi, A.D. 1995. Ecofriendly Botanical Pesticides. Intensive agriculture 37(1-2): 37-38.

Khan, M.J.2010 Changing Dynamics: Crop protection and Indian Pesticides Industry Agriculture Today 13(12).34

Sharma, A.2022. Transition from Conventional to Organic Farming. The Daily Excelsior. 58(90):8.

Singh and Arvind Prakash, 2022. Vegetables for all. Personnel Communication on the address of Prof. (Dr.) R.D. Gupta. 258-C, Sainik Colony, Jammu-180011, J&K. Today.

Yadav, MP. 2010. Organic Farming Bone or Boon for food security. Agriculture.Today.13(12):9

8

Role of Rural Women in Sustainable Agriculture through Organic Farming

Agriculture is the main stay of Indian economy as more than 70 per cent of the people of India are dependent upon this avocation and allied disciplines for their livelihood. Even today, at the dawn of 21st century, the Indian agriculture has continued to be the backbone of our national economy, representing country's one third of Gross Domestic Product (GDP). Thus, farming families consisting of men, women, boys and girls known as the pillars of our agriculture, constitute backbone of the nation. The support and contributions made by the rural women in agriculture, constituting 55-66 percent of farm work force (Singh et al., 2012) are, however, not less than those performed by the men, and therefore they become spine of the rural economy. In fact, women's role in the economy has been underestimated and their work in agriculture has long been invisible and unappreciated, if we compare the women's work participation rate of some of the selected countries like Indonesia, Bangladesh, China. Pakistan and Sri Lanka, we find that India ranks second in position (22.7 %) after Indonesia (33.4%).

In India nearly 74 per cent of the entire female working force is engaged in agricultural operations i.e., 28 per cent as cultivators and 48 per cent as agricultural labours. Generally, 60-70 per cent of labour input is provided by women which increases to 80 per cent during agricultural peak season.

The historians believe that it was the women who, for the first time, started growing of crops and domesticating animals and thereby, an art and science of agriculture farming initiated These days about half of the available global human resources were women (Singh 2004, Singh 2012). The women, contributing to agricultural production are usually called as "Peasant women" i.e., those women who work in the field but are not the wives of the farmers. On the other hand, those women who are the wives of the farmers are called "Farm women". On an average, farm women work for longer hours than the men i.e., 69 hours per week as opposed to 57 hours per week for men. Most of this is classified as housework which is unpaid or is subsistence levels economic activity, and brings little remuneration. Then there is the system of "Patriarchy" inheritance whereby daughters, sisters, wives and widows, are entitled to smaller shares of property and decision making (Kheterpaul & Grover, 2001).

Although India has achieved self-sufficiency in food grain production yet its sustainable supply is a challenge in the light of the burgeoning human population. In developing countries, agricultural production is mainly dependent upon manual labour, the lion's share of which comes from the rural women. This situation also holds well in India, where women provide 75 per cent of the agricultural labour. Nearly, 84 per cent of all economically active women in India are engaged in agriculture and allied activities. Roughly, they constitute one-third of the agricultural labour force and 48 per cent of self-employed farmers. It is worth mentioning that there are 75 million women as against 15 million men in dairying, while the women engaged in animal husbandry account for 20 million as against 1.5 million men. Not only this, establishments of kitchen gardening, laying out of lawns, and managing of floriculture, are handled by the women efficiently. The main focus of the chapter is to highlight women's role in "Sustainable Agriculture".

Women and Origin of Agriculture

History gives evidence that it was the women who started to raise or grow crop plants about 12000 years ago and, thereafter, an art and science of farming or agriculture came into existence. Again, it is believed that it was the women who, for the first time began to domesticate animals about 25,000 years ago (Gupta and Sharma 2002). Thereafter, science of rearing of cattle i.e., animal husbandry came into being.

It is accepted as true that while men used to go out for haunting in search of food. The women used to collect seeds from the existing flora and set out cultivating them for obtaining food, fodder, fuel and fibre, which all make up the basic necessities of life. In this way the women have on umbilical attachment with agriculture since time immemorial.

Contribution of Women in Food Grain Production

Rural areas are most often associated with farm and home activities. The contribution of the farm women in enhancing food grain production has a special significance could be judged in terms of influencing the farmers with respect to the following technologies:

i. Accepting and adopting new technologies in agriculture including, floriculture and animal husbandry to increase their production and, thereby, to enhance income of the farmers.

ii. Modernizing the farms through improved implements, tools and other farm machinery.

iii. Developing the farms with irrigation facilities depending upon the availability of irrigation sources.

The roles of men in performing such activities are numerically insignificant. In other activities like thinning, gap filling, weeding, hoeing, harvesting and threshing vegetable, picking, processing and storage, the farmwomen share the work with men folk.

Livestock husbandry is another important activity, which can be looked after very nicely and efficiently by the women. It has been found that women can manage the dairy animals and small-scale poultry better than men. Women can very easily look after up to 3 cross bred cows in addition to taking care of her own family. From these cows, she can earn an annual income of Rs. 15000-18000.

In livestock husbandry or livestock farming, the women are actually involved in operations like feeding, breeding, management, health care and marketing of animals including green-fodder production. It has been observed that in dairying 75 million women are involved as against 15 million men.

These days, the participation of the rural women has become an utmost important in decision making, processes relating to various farm activities Further an involvement of rural women in agriculture has been felt to make them knowledgeable or create awareness about the ill effects produced by the "Green Revolution" and tackling them effectively. For example, the rise of agro-chemical inputs in developing countries under Green Revolution and the migration of male heads of households from farmlands to cities, in search of urban jobs, leaving the women to take on an increased burden of agricultural operation. This has resulted in getting the women exposed to toxic chemicals in increasing numbers leading to endocrine disruption related health problems. Endocrine disruptions, the World Watch Institute's State of the World 2002 reports note, "is a far more serious health problem than cancer". Agrochemicals have also been associated with pregnancy failures and infant developmental problems.

It also becomes imperative to make the women familiar with the latest agricultural technology like role of biotechnology, use of biopesticides;

biofertilizers and vermicompost, concept of integrated pest management and integrated nutrient management, which all are essential for sustainable agriculture.

Sustainable Agriculture

a) Women and Basic Life Support System

Rural women are playing and will play a pivotal role in conserving/preserving, basic life support system such as land, soil, water flora and fauna. In rural areas, they usually get engaged in farm and home activities. "The nature and extent of women's involvement in farm activities, however, varies widely not only from region to region but also within the same region, depending upon the differences in the ecological subzones, castes and classes, stages in the farm cycle and farming systems. Hardly, there is any activity in agriculture of food production, except heavy physical labour like soil excavation or ploughing the fields wherein women are not actively involved. In fact, the rural women are able to perform a number of operations more efficiently than performed by the men.

b) The Myth of the Green Revolution

Most of this mayhem has been carried out in the name of food production. Although the Green Revolution increased agricultural production, yet it required higher and higher inputs to get the same yield. For example, according to the Economics and Statistical Departments of Punjab, in Ludhiana District during 1970-1971, 853 Kg fertilizer ha^{-1}. Ten years later, during 1981-1982, applying 167.2 Kg fertilizer gave 33.79 q yield $ha^{-1}$of some crops. Thereafter, without balanced use of NPK fertilizers, one could not obtain continuously sustained high yields without detriment to soil. Increased use of farm yard manure along with NPK fertilizers was found suitable. Use of nitrogenous fertilizers alone has started giving negative effect and proved most disastrous in low pH soils (Kanwar, 1997). Moreover, the deficiency of micronutrients like Zn, Fe, Mn, Cu has become a yield limiting factor in many soils.

Millions of hectares of once productive farmland have gone, and continue to go, out of production. While elite farmers who could manage their farms scientifically were able to optimize the use of inputs like improved varieties, irrigation facilities and agro chemical for economic crop production. However, a large number of small holders have ended up with miserable failures. There are a number of examples, where farmers have lost their soil productivity and crop yields due to excessive use of water for irrigating the crops or improper use of agrochemicals, particularly crops like cotton. Indiscriminate use of chemicals has resulted in reduction in the biodiversity of natural enemies, outbreak of secondary pests, development of resistance to pesticides and contamination of food and the ecosystem. Above all without reliable lean season water supplies and stable soils our food security is unquestionably at risk. All this reaffirms that intensive use of external inputs may not be good under all situations.

c) Agriculture Models Devised by Women

In case of small and marginal farmers, particularly in backward areas or in arid zones, it is necessary to encourage them to build up their farming operations gradually with very little use of external inputs. Here, the rural women can help in devising some agricultural models, which can sustain agricultural production and productivity. The women have less working capacity than men can do. Hence, they devise small and efficient agricultural models with more sustainability due to having traditional knowledge and skill of various farm operations. For example, "Bara Anaj" the typical mixed cropping system of Uttaranchal hills has been evolved by the rural women. In this farming system, almost all cereals and pulses required in a household, are cultivated in the same field, hence the term "Bara Anaj". This is totally self-sufficient sustainable agricultural practice, preferred by the women in the hills. Thus, keeping in view the traditional knowledge and skill of farmwomen in different farm activities, it will not be wise to ignore the possibility of increasing avenues, which uphold the country's food status through utilization of this unattended and unrecognized force.

In the north eastern hill region, covering the states of Meghalaya, Nagaland, Manipur, Tripura, Arunachal Pradesh. Sikkim, Assam and Mizoram, there exist uneconomically viable and sustainable indigenous farming systems, which were developed by the tribal farmers including farmwomen through their ingenuity (power at creative imagination) and skill "Zabo" farming system and "Apatani" farming, are some of the examples. In Zabo farming system, chakheson tribal tarmus in Phek district (Nagaland) have combined the forestry, agriculture and animal husbandry with soil & water conservation. In "Apatani" farming system, Apatani tribal farmers have devised rice based fanning system combined with finger millet production and fish culture.

d) Close Linkage of the Women in Sustaining Food Cycle

In sustaining agriculture based on maintaining the soundness, health and fertility of soils, the women have played and continue to play a paramount role, particularly providing a close linkage in sustaining the food cycle. They assist in feeding animals, collecting foliage from bushes and trees, gathering crop by-products and cutting the grasses from grasslands. They prepare farmyard manure and compost and fertilize the fields, with these manures. This protects the health and fertility of the soils and, thereby their productivity. In fact, agriculture based on women's participation is natural, self-reproductive and sustainable as the internally recycled sources provide the necessary inputs like good quality seeds, soil nutrients moisture and biopesticide for pest control. This system of agriculture does not harm the environment, rather maintains ecological balance. In reality, it looks for an ecofriendly approach for evolving and popularizing ecology low cost input and environmentally sustainable basis.

e) Farm Women and Crop Diversification

The farm women give their full help and cooperation in managing crop diversification methods. Such methods mostly consist of crop rotation, mixed cropping and multiple cropping. The main benefits of diversified agriculture include to check soil erosion, improvement in soil fertility and enhancement of crop yield. This also reduces the need for nitrogenous fertilizer in case of legumes. In crop failure due to the pests and diseases, where even if one crop fails there remains other crop in the field that with stands yield and supports the farmer.

Genetic diversity, location specific varieties of various crops and use of local plant species, shrubs and tree species as an inexpensive source of great manure to build up or maintain soil organic matter and fertility are essential for realizing sustainability The rural women have proved boon in collecting these materials for serving as green manures. In areas of India, which are still free from Green Revolution, the farm women thus continue to work there as soil builders rather than soil predators. It is from these areas or tracts of natural farming, are emerging the ecological struggles to protect the nature.

Horticulture requires intensive and tender care, which can be handled efficiently by women. Establishment of kitchen gardens, nurseries, flowering gardens, laying out of lawns with growing of medicinal plants, vegetables and fruit trees in addition to grasses, are now totally managed by the rural as well as urban women.

In Vausda Taluka of Valsad district in Gujarat, BAIF Development Research Foundation, Pune, a voluntary research organization has promoted a nursery of mango and cashew grafts through tribal women. Each group of 4 -5 women have abled to raise about 5000 plants and earn a net profit of Rs. 6000-8000 every year.

Apiculture (Beekeeping) is another area which is required to be increased by way of increasing the number of bee colonies and covering more area under it. It is stressed upon that bee keeping can play a bigger or major role in reviving even horticulture due to cross pollination. The officers of agriculture department are required to devise more brands of honey based on the feeding flora. It is also suggested to repacking and redevising market strategies for honey and even asked them to explore new market segments like airline etc. Apart from this, the farmers are asked to take up mushroom cultivation in a big way and must provide market avenues for the same.

It is inferred that women's participation in economically agricultural productive activities is not new but it is a time immemorial practice (Gupta, 2015). Now it is a matter of common observation that the women not only engage themselves in many of farm operations in their own fields but also work in the fields of others as hired labours.

Women and Forest Gathering Activities

Famine has not been a problem in the tribal areas of Bastar (Madhya Pradesh, India) as the tribes have been able to obtain half of their food from innumerable edible forest products. These tribals use about 165 trees, shrubs and climbers. Out of these trees, there are 31 plants, the seeds of which are roasted and eaten by these tribals, 19 plants whose roots and tubers are eaten after baking, boiling or processing, 17 whose juice is taken, 25 whose leaves are taken as vegetables and 65 plants whose fruits are taken raw, ripe or roasted and picked. They also constitute an important source of dry season fodder. Forest-based gathering and processing enterprises provide seasonal employment and sources of income Women often dominate forest gathering activities, both for household products and income. For example, collection of minor forest produce, particularly oilseeds like neem, mahuva, pongmia, sal is an area where women play a significant role. It has been estimated that neem seeds worth Rs. 1000/- million go waste every year. The main reason for such neglect is lack of awareness about their values and uses as well as inadequate facilities for processing and marketing. If these issues are tackled, oilseed collection is an attractive item for rural women to generate additional income thereby boosting significantly country's GDP.

Oil (Margosa) extracted from neem seeds is useful and powerful anthelminthic. As farmers of the hilly areas of Jammu region, especially of Poonch, Rajouri, Udhampur, Kishtwar, Doda and Reasi districts have animal husbandry an important sector in their agriculture, so animal husbandry and sheep husbandry departments are required to render valuable service to the livestock breeding in respect of both health as well breeding cover. The curative health cover should also be provided by way of doing vaccination against various diseases. The departments are also required to provide health coverage facilities to the migrant's livestock which moves to highland pastures during summer season.

Women's participating in economically agricultural productive activities is not new but it is a time immemorial practice (Gupta, 2013). Now it is a matter of common observation that the women not only engage themselves in many of the farm operations in their own fields but also work in the fields of others as hired labourers. It is believed that women's share of labour hours has increased disproportionately to that of men. Women, infact, constitute almost half of the agricultural labour force, especially in organic farming and contribute more towards family income.

References

Anonymous. 2015. Boasting India's Livestock Agriculture Today. 18 (6): 20-27

Anonymous. 2015. Livestock, Dairy, Poultry and Fisheries- An overview Agriculture Today.18(6):28-34.

Anonymous. 2015. Malra goat can produce Pashmina worth Rs 20 crore annually: Study. The Tribune. 135(165):5[Friday, 19 June]

Gupta, R.D.2009. Pashmina goats in Ladakh- Origin and rearing. Glimpses of North West

Gupta, R.D. 2013. Farm women and future strategies. The Daily Excelsior. 49(361): 6

Gupta, R.D. and Sharma, A.K. 2002. Organic farming in Jammu and Kashmir. The Daily Excelsior 38(330):6

Himalayas. Radha Krishanan and Co. Pacca Danga Jammu Tawi.180001, J&K, (India). Kanwar, J.S. 1997. Fertilizer Policy Issues (2000-2003) .National 3rd Agricultural Science Congress.

Singh, S. Khan, M., Shukla, U. N and Sharma, A.2012. Role of women in agriculture.Agrobios Newsletter.11(3): 71-72

Singh, S.K. 2004. Organic farming a key issue for Indian Agriculture Indian Farmer 's Digest 37(2): 5-6

Singh, Yogendra. 2012. Biological aspects of organic farming for sustainable agriculture. Indian Farmers Digest. 45(5): 35-37.

9

Organic Rose Cultivation

Rose (*Rosa centifolia* Linn) belongs to the family Rosaceae is known as flower of love which is cultivated throughout the world. The traditionally grown rose flowers are fragile wild, bloom only for a limited period. They are found in a few colours of red, pink, deep crimson vis-a-vis white colour. However, with the development of the clonal technology, tissue culture and hybridization, a wide variety of roses can be grown having different colours, fragrances, blooms and shapes. It is point to mention that from the 200 species of roses originally found there are more than 30,000 varieties available these days.

The cultivation of rose is believed to have started about 5.000 years ago in China. Initially, Romans started to cultivate the rose flower using organic manures for medicinal purposes and then tor aesthetic values. Briefly their medicinal values have been mentioned separately as reported by Gupta *et al* 2016. It is not until late 8th century that cultivation of roses was introduced into Europe. Unlike Europe and America where rosses have limited flowering period during the summer.

In India, though several rose species were found in the Himalayas, its cultivation started much later. In India, especially in its northern parts the true rose season is from the end of November to end of March and some varieties of rose bloom during June and July with good fragrance. Roses, however, grown in India are often attacked by pests which must be treated organically. The inorganic or chemical pesticides used on them have proved very dangerous for the health of the human beings especially the babies (Hathi,2011).

Use of Organic Pesticides

1) **Use of Tomato Juice:** Make a solution of tomato leaves; add four to five pints (1/8th of gallon) of water and a table spoon of corn starch, strain it and then spray this on rose flowers grown. One can even just grow tomato plants around the roses, and they will become safe from the black spot.

2) **Use of Baking Soda:** This is another method that had been tried and is based on the research done by Dr. R.K. Horst, Professor of Plant Pathology at Cornel University, United States of America (USA). In this method, a 0.5 per cent solution of baking soda (sodium bicarbonate or bicarbonate of sodium ($NaHCO_3$) per gallon (3.7 litres) of water is sprayed on roses. This is a good remedy for tackling the black spot and powdery mildew.

3) **Growing of Chives:** Grow chives around the rose flowers grown. They are available in many nurseries now. Organic rose gardeners who have used chives, have never found aphids attack in their rose flowers grown again.

4) **Hand Picking of Pests:** Handpicking the pests is another good option. While handpicking the pests, one must remove and destroy the leaves attacked by the aphids to reduce their recurring.

5) **Use of Strong Spray of Water:** Another simple and safe method is to use a strong spray of pressurized water to knock aphids of the roses.

6) **Use of Companion Plants:** Companion plants here refer to grow plants like garlic, coriander and petunias around the rose flowers grown. Growing of such plants will definitely help to check the pests of rose plants. If spider mites attack roses, they suck moisture from the plants and suffocate them in a web. To control such an attack is just to wash the mites off rose plants with a strong spray of water.

7) **Growing of Garlic and Onion Plants:** Grow garlic and onion plants around the rose plants. One clove of garlic plant near the roses will repel aphids and green flies. Garlic also exudes sulphur which will definitely kill black spot fungus.

8) **Use of Insecticidal Garlic Spray:** There is also an insecticidal garlic spray that one can use. Soak three ounces (0.08/Kg) finely minced garlic in two teaspoons mineral oil for 24 hours slowly add one pint water that has been mixed with one forth ounce (0.089Kg) insecticidal soap. Stir thoroughly and strain into a glass jar for storage. One can use one to two table spoon per pint of water for spraying on roses. Sometimes this strong spray may cause leaf damage of rose flower. The solution is then diluted a little.

9) **Use of Eggshells:** Collect egg shells, crush them and put a thin layer of them around the rose plant stems. Then cover the egg shells with soil. There will be two benefits - eggshells will provide calcium to the soil

and they will repel root maggots and cut worms that are present there. As eggshells are sharp and soft little insects will find it uncomfortable to crawl through them.

Main Organic Rose Gardens

Roses now range from simple exhibition red to breath takingly gorgeous exhibition type flowers with a spectrum of colours. Now a days, a number of rose gardens have been prepared where rose flowers are mostly grown organically.

i) **Rose Garden at Chandigarh:** Rose Garden sprawling over an area of 15 acres (6th hectares) in Chandigarh was established in 1967. It houses one of the finest varieties of rose flower numbering about 594.

ii) **National Rose Garden Delhi:** It is situated in the Embassy area. Chanakyapuri of South Delhi. Large varieties of rose flowers are grown in this garden under the premises of society of India.

iii) **Government of Rose Garden, Ooty:** Earlier known as Jayalalitha Rose Garden, is situated on the slopes, of the Elk Hill in Ooty (Tamil Nadu) and is positioned at a height of 2200 meters. This garden was set up to memorialize the anniversary of the Flowers Show held in May 1995. It is one of the greatest tourist attractions of Ooty in Tamil Nadu. It sprawls over an area of 10 acre (4.04 hectares) of land. The rose flowers are positioned in five twisted terraces. One can be the best, the greatest collections of roses like mini roses, hybrid tea roses, ramblers, black-green roses and many others of unique varieties.

iv) **Medicinal values of Roses:** Rose not only represents beauty, love and affection but has lot of medicinal values. It makes the heart and mind cheerful. Wash the eyes with rose water to reduce the redness of the eyes as well as their swelling. The gulkand made out of rose is purgative in nature and reduces the heat in the intestine and stomach. A few drops of rose essence used in sweet dishes to provide coolness.

Rose extracts as well as its oils, are well known for their emollient and hydrating properties, assisting the skin of human bodies to stay soft and supple (Gupta et al. 2016). The moisture present in rose flowers possesses a bit of tannin which serves as skin tonic.The rose oils add radiance to complexion, whereas rose extracts are rich in vitamins A, C, D and E as well as flavonoids. Flavonoids serve as antioxidants for the skin. The stimulating action in rose petals aids to fight early ageing, enabling a soft and youthful complexion.

Jangli gulab (*Rosa moschata*) is a climber in Chirpine (*Pinus roxburghii*) forests and deodar *Cedrus deodara* forests occurring frequently has not been investigated scientifically. Locally it is used in eye trouble such as washing of sore eyes with its decoction.

Another species belonging to the family the *Rosa macrophyll* as a hardy plant and is in flower from the Month of June to July and seeds ripen from August to September. The edible parts of this plant are the fruit and the seeds. The fruit, whether raw or cooked, is said to be very sweet when bletted (eaten when overripe but before it starts to rot). The bottle shaped fruit is about 4 cm long, though it can be up to 7 cm long but there is only this layer of flesh surrounding the many seed. Some care has to be taken while eating this fruit (Jamwal, 2009) as there is a layer of hair, around the seeds just below flesh of the fruit. The hair can cause irritation to the mouth and digestive tract if ingested. This point requires to be repeated so that those eating the fruit are careful, because of the main hazards while eating this fruit.

The seeds are a good source of Vitamin E. These can be ground and mixed with flour or added to other food as a supplement. Be sure to remove the seed hairs.

The list of medicinal uses identified includes the treatment of Cancer and for Ophthalmic purposes. A paste of the fruit is taken with the belief that it is beneficial for the eye sight. The fruit of this genus is a very rich source of the vitamins and minerals especially in vitamins A,C and E as well is a rich source of flavonoids and other bioactive compounds. It is also fairly good source of essential fatty acids. It has investigated that fruit serves as a good food for reducing the incidence of cancer (Jamwal, 2009).

Edible roses belong to the same species as nonedible ones, but fertilizers have to be organic ones (Anonymous, 2016).

Other Uses: It can also be used or utilized as a decorative or separating/ screening hedge.

References

Anonymous. 2016 Edible roses Tribune. Sunday [28.9.2016]

Gupta. R.D., Gupta S.K. Bhardwaj. S.D. 2016. Agrotechniques and uses of Medicinal Plants - Associated Publishing Company. A division of Astral International Pvt. Ltd. Nee- Delhi,

Hathi. D. 2011. Go Organic on roses. The Tribune, 131 (119): 3 [Sunday spectrum].

Jamwal, J.2009. Rosa macrophylla. The Kashmir Times, Jammu. 66 (189):13 [Thursday, July March 6]

Mohindra. K.2010. Floral remedies. Ayurveda for Holistic Health. 1(17): 14

10

Organic Pulse Farming

Pulses are an important source of proteins. One hundred grams of pulses contain about 25-32 grams of proteins and several amino acids which are not made by the human body (Sidhu, 2015). These are the staple food both for vegetarians and non-vegetarians of the country. India is not only the largest consumer of the pulses in the world but it is also the largest producer and importer. The global production of pulses is about 70 million tonnes, of which India's share is around 25 percent.

The consumption of pulses is historically part and parcel of Indian culture. However, over a period of time, the availability of pulses has declined from69 gram per capita per day in 1961 to 51 gram in 1971 and 38grams in 1981. Marginally it arose to 42 gram in 1991 which already declined to 30 gram in 2001. The National Food Security Mission (NFSM) was started by the union Government for pulses during 2007. It did not result in a significant increase in the output because pulses are grown only in 16 per cent area which is irrigated. The area under cultivation for pulses varied from 23 million hectare in 2009-2010 and 26 million hectares 2021-2022 and production was14.6 mt. to 28.9 mt. At present the availability of pulses is about 42 g per capita per day.

India imported 13.75 million tonnes(mt.) pulses in 2009-11 to 27.71 million tonnes in 2021-22. The average import per annum was worked out 3.0 million tonnes out of the total domestic supply. The imports included 1.95 million tonnes of yellow/green peas and 0.82 million tonnes of lentil mostly from Canada and USA,62 million tonnes moong and urad from Myanmar, 0.58 million tonnes of tur(arhar) from Myanmar, Tanzania and Mozambique and

0.42 million tonnes chana from Australia and Russia. Since pulses production declined from 19.8 million tonnes in2013-14 to17.4 million tonnes in 2014-15, it had adverse impact on the domestic availability in spite of import of 1.6 million tonnes of pulses in the later year. It is point to mention here that we have faced two successive deficient rain falls in 2014 and 2019 witnessed un seasonable increase of Rs 32 to Rs. 50 per Kg due to successive rains in March 2015 in the northern region. Climate has become big issue as for as agricultural sector all over the world is concerned because it has resulted in adverse effect on the farm outputs including pulses.

The Union Government alone is not responsible for the spike in prices of pulses as witnessed due to mismatch between the force of demand and supply of pulses. There are multiple factors which are behind the crises. The International prices of pulses have increased from Rs32 to Rs 50 per Kg, Rs56-Rs75 per Kg for lentil, Rs40 to Rs 90 per Kg for tur, and Rs50 to Rs. 77 per Kg for urad between Oct.2014 and August 2015.The high import prices have an impact on the domestic wholesale and retail prices of the pulses.

The measures against the hoarders had have some impact and retail prices have come down marginally. The Union Government's decision to import pulses has also had an impact on the retail prices. In agriculturally developed states like Punjab, Haryana and Western U.P, the area under pulses declined drastically in the post Green Revolution period due to more emphasis on wheat and rice cultivation by the Government. These two principal cereals were covered under the public procurement with minimum support price since 1966. The yield of wheat and rice is more stable as well as high and therefore, it provided higher income return as compared to pulses.

Under the diversification plan of agriculture in Punjab, there is a proposal to bring additional 70000-hectare area under pulses by the year 2017-2020. Madhya Pradesh is the biggest producer of pulses in India with about 20% share in the production followed by Maharashtra (16%), Rajasthan (19%) and U.P(9%).Among different pulses, the gram is dominant one with around 45% in production. It is followed by 15% tur,10% each of moong and urd,5% of lentil and 15% other pulses like peas, moth etc. The long-term solution of the pulses crises lies in increasing productivity and production because there is little scope to increase the overall area and cultivation pulses in the country. Cultivation of pulses can be promoted in agriculturally developed states like Punjab and Haryana where assured irrigation is available.

In the overall interest of Indian agriculture, the production of pulses can be promoted on the pattern of rice and wheat. A buffer stock of one to two million tonnes of pulses can be built up in the future. The recent decision of the Union Government to have a buffer stock of 40000 tonnes of pulses is to meagre because the consumption of pulses is higher. The policy planners. Agricultural scientists,millers,traders and farmers join hands and work together to make the country self sufficient in pulses production during the next seven to eight years. Success attained with cereals particularly rice and wheat can be replicated with pulses. Prime Minister, Jawaharlal Nehru in a meeting with the faculty of

IARI,New Delhi in 1947 said, "Everything else wait but not agriculture." This advice can be taken seriously.

India has the distinction of being the largest consumer of pulses in the world(Mahajan,2015)'Prices of the pulses have been rising for the last so many years and now they have beyond the reach of the poor and have increased by more than 40% during the year 2014.With the prices of pulses going upward, the centre has hurriedly ordered the import of the pulses and also created a fund of Rs 500 crores to pay for the transport and processing of the pulses thus imported.

Insufficient Production

It is notable that the demand for pulses has increased significant by due to population increase. There was no much expansion in the area also. In 1970- 71 productivity in pulses was 524 kg per hectare which increased 785 kg in 2013-2014 and 885 kg/ha in 2020 whereas in case of wheat per hectare productivity increased from 910 kg to 3030 kg during the same period i.e. there was a 3.6 times increase. Thus, decline in per capita availability of pulses posed a serious challenge to the poor man's food basket. The most common method to tackle the shortage in any commodity is to import that commodity. Pulses are no exception to this rule. We have been importing edible oils also besides pulses. In 2014-2015, we imported pulses worth 2.8 billion dollars. The international price of moong is nearly rupee 80 per kg now. If we add the transportation and handling charges, the price of imported pulses is much higher. There is a dire need to make pulse available so as to fulfil the protein requirements, according to people's preference and food habits. To fulfil the requirement of 130 crore people in the country, we cannot depend always upon import. It is therefore imperative to raise production of the pulses in the country. History tells us that assured minimum price in pulses also as was done in case of wheat and rice.

This way we can promote production of pulses. It is notable that international price for tur is rupee 80 per kg and the government even purchased the same, in view of shortage in such as case the promotion of domestic production of pulses will go long way in improving the conditions of the farmer by following the below mentioned methods.

1. There is a dire need to grow pulses organically using organic manure and rhizobium botanical culture wherever it become necessity.
2. Provide the better quality of seeds to the farmers

3. Create irrigation facilities wherever it become possible.

Vice president Mohammed Hamid Ansari today said the central proposal to push agriculture through a second green revolution could start from Jammu and Kashmir basis on pilot basis.

'It is agriculture that has potential to revitalize the economy of the state and also bring about greater social inclusion and sustainable rehabilitation'. SKUAST and Governor Lt General (Rtd.) S.K Sinha conferred Degrees to the students

Fall in Farm Output

India is primarily an agricultural country and about three fifths (about 60%) of her population resides in villages. Majority of the people in the villages depend up on agriculture for their livelihood. But unfortunately, our agriculture is rainfed and it depends up on monsoon rains which play a vital role in crop production. Good monsoon rains promise us as bumper crops whereas failure of monsoon leads to fall in crop output. It really said that our country has been facing severe climate conditions like floods, drought and failure of rain leads to fall in farm out. Harvest of crops affects the farming community adversely and hey find it difficult to make both and meet such an opportunity to the agriculture scientists to conduct research and develop such varieties of crops including pulses which can tolerate the climatic fluctuations. Now a days, change of climate has also come to force.

Thus, institutions such as IARI and many others have come to play a key role to develop of climate resilient technological solutions in live rating opportunities from science such as biotechnology, nanotechnology synthetic biology etc. The said Institute has already developed mustard which can be in uncultivable area, our education in agriculture too needs to be focused on developing pulse varieties the seeds of which become to Indian climatic conditions.

Awareness camps too need to be organised for the farmers to enlighten them about various schemes such as crop Insurance, use seeds of new varieties of pulses, proper use of pesticides, manures and fertilizers. Efforts must also be made to promote organic farming in India to improve rural economy through application of biofertilizers as well as use of biopesticides. While planting or sowing of pulses the farmers must be made fully acquainted with the use of rhizobium inoculant or rhizobium culture. This inoculant or culture is a kind or type of biofertilizer. Rhizobium inoculation significantly enhances the pod number of black gram, green gram and red gram as well as their grain weight. Their increase in grain weights and, thus, thousands of grains weight is mainly responsible for higher grain and straw yields.

This is mainly attributed to continuous supply of nitrogen to the plants through Rhizobia - Symbiotic bacteria confined to legume modules of root, having capacity to fix atmospheric nitrogen. This effect is further enhanced by using rhizobium inoculation. Thus, sowing of pulses will not merely supply of proteins to the human beings or other animals but also supply atmospheric nitrogen to soils.

As a matter or of fact, use of biofertilizer especially use of rhizobium cultures in pulses, are found extremely beneficial in enriching for fertility of the soil and carry out plant nutrients requirements by way of providing nutrients organically via microorganisms and their own by products (Singh and Sadawarti, 2020).

What are Rhizobia?

Rhizobia are "Gram Negative Bacteria", which are generally referred to as the root nodule bacteria. These bacteria live in root cells of pulses being the leguminous plants where their growth and metabolic activities cause a swelling to form modules on their roots through infection thread (Gupta, 2007).

How Rhizobium Bacteria Function

Rhizobia are capable of utilizing the free atmospheric nitrogen and synthesizing it into various nitrogenous compounds. Pulses being leguminous plants obtain their nitrogen from the synthesized nitrogenous compounding while the Rhizobia derive their energy from these plants. Pulses like clover, sweet clover, alfa-alfa, vetch, pea, bear, lentil, gram, soybean etc. are prominent among the leguminous crops. As already stated, that one of the characteristics of these plants is their ability to live in mutual association with rhizobia which render or make nodules on their roots. Thus, the pulses like other leguminous plants, and the rhizobia live together from the mutual benefits of each other. Such an association is known as "Symbiosis" (Gupta, 2007; Pareek, 2014). As, the rhizobia fix atmospheric nitrogen in the root nodules of the pulses and other legumes as such there is no need to add nitrogenous fertilizers to these plants.

Types of Nodules

Like other leguminous crops, pulses are also characterised by the presence of two kinds of nodules viz; (i) Effective nodules and (ii) Ineffective nodules. The fact, whether a rhizobium legume association is effective (capable of fixing atmospheric nitrogen effectively) can be ascertained even in the fields by uprooting the legumes and examining certain characteristics. For example, the effective modules are mostly pink or brown in colour with reddish pink inside whereas ineffective nodules are usually whitish or greenish in colour. It is point to mention that pink or brownish colour of the effective nodules is due to the presence of Leghaemoglobin which is quite similar to Haemoglobin that is found in human's blood. Presence of leghaemoglobin in effective nodules, is responsible for fixing atmospheric nitrogen.

Contribution of Pulses

Pulses are the integral part of the Indian agriculture and contribute substantial amount of nitrogen to soil. However, since they are generally grown on marginal lands with poor crop management, resulting in a situation that sometimes they deplete soil nitrogen (Pareek, 2014). There are, however, a number of issues relating, to biological nitrogen fixation in pulse crops which can be rectified by following the below mentioned strategies (Mahajan, 2015).

1) The first and the foremost strategy which requires to be followed by the farmers is to always apply organic manure in the fields before sowing pulses It is because a soil capable of supporting sustainable pulses or agriculture production must contain all the essential nutrients in available form and proper proportion, which can be resorted only through organic manures. Addition of manures is important not only for maintaining supply of plant nutrients but also for buffering action of soils, by way of improving soil structure, preventing soil Frost and more often than not, decreasing the incidence of plant diseases. Application of organic also improve biological properties of Soils too (Gupta, 2013).

2) Another strategy is to apply Rhizobium inoculants. The fundamental purpose of Rhizobial inoculation is to add a fresh Rhizobial culture of effective strains of legume bacteria to soil or to the seeds of the pulse crop. Effective inoculation of legumes bacteria to soil or to the seeds. Effective major inoculation of legumes has been a factor in improving their yield and productivity good growth of the pulses like other legumes contributes to the maintenance of good matter. Well decomposed quality of organic matter be, humus in turn improves physico-chemical properties of the soils. It increases the moisture holding capacity and cation exchange capacity and helps plant nutrients.

As early as 1896, the legume Rhizobium symbiosis was found to be the potential source for exploitation of Biological Nitrogen Fixation (BNF). This later on resulted in the development of legume inoculants and use of artificial inoculants of various legumes at the time of sowing of legumes including pulses at the time of sowing of legumes in order to enhance their modulation by way of using efficient strains and in return to increase their yields. Now there are many reports which have indicated that there is substantial increase in the yield of various legumes including pulses through the use of Rhizobial inoculants. The response of different legumes grown in various parts of the country to Rhizobial inoculants has given promising and results. However, the response and performance of a particular Rhizobial strain varies depending upon the soils and agro- climatic conditions of the region.

Factors Responsible / Affecting Nitrogen

Amount of atmospheric nitrogen fixed by various legumes including pulses depends up on a number of factors. Some of them which are the most important, are as follow:

1) Kind of legume or pulse
2) Effectiveness of the pulse legume
3) Type or condition of the soil
4) Presence of essential plant nutrients exclusive of nitrogen

It is worthwhile to mention that in the highly fertile soils supplied with more available nitrogen, there is no atmospheric nitrogen fixation occurs. Most noticeable results from legume inoculation are obtained in soils having low or average soil fertility.

Amount of nitrogen fixation by various pulse crops: Nitrogen fixation by various pulses, is shown in Table 2. The data shown in the said table indicated that red gram and soybean are capable to fix the highest amount, of nitrogen fixation in the range of 168-200 kg ha^{-1} and 40-256 kg ha^{-1} respectively followed by the and peas in the range of 85-110 kg ha^{-1} and 51-132 kg ha^{-1}, respectively annually.

Apart from the above, growing of pulses in hilly area like cow pea, soybean and other types beans render lower soil erosion losses as compared to growing of non-legumes like pearl millet, sorghum and maize, etc (Gupta and Sharma, 2007).

Biochemistry Relating to Biological Nitrogen Fixation

Various aspects of biochemistry relating to biological nitrogen fixation as reviewed by Smith (1978) are as under:

1) Studies on the mechanism of nitrogenase action indicate that acetylene and dinitrogen are reduced by different forms of enzyme and after a biochemical rationale for variations in the reduction ratio.
2) Evidence on the nature of the nitrogenase active site is reviewed.

3) ATP has more than one role in nitrogenase action and the ATP/ADP ratio is probably important in control of nitrogenase activity.

4) There is evidence for the involvement of glutamine synthase, oxygen and molybdenum in the regulation of nitrogenase synthesis.

In the end, it is concluded that the soil quality is the basic pillar for maintaining the sustainability in agricultural production including growing of pulses (Sabharwal, 2022).

Of the ancillary proteins associated with nitrogen fixation: (a) the electron carrier to nitrogenase is variable but require an oxidation-reduction potential of less than - 400 mv; (b) leghaemoglobin in the legume rhizobium symbioses, facilitate oxygen diffusion to oxidase sites which can then generate ATP for nitrogenase action.

Salient Features of Pulses Cultivation

About of the 80-90 per cent area under pulses in India is non irrigated. The crops are deep rooted and have low irrigation requirements. The most commonly grown pulses can be divided into two main categories:

a) Those which are sown during rabi season after the cessation of monsoon rains and grow under the decreasing soil moisture.

b) Those pulses which are grown in the rainy season and are subject to conditions of excess moisture or deficient moisture at different times during the crop season. In former category are included crops such as grams, peas and lentil. The important crops included in the latter are black gram, green gram, guar, cowpea, moth. Red gram is long duration crop i.e; legumes which are grown with the commencement of the monsoon rains and mature during rainy season.

It is remarkable to note that pulses may be grown as pure or mixed crop with other crops.The pulses grown in the kharif season are sown as mixture with jowar,rice,maize,and cotton.Rabi pulses are mixed with wheat,barley,linseed and rapeseed and mustard. Growing of leguminous crops including pulses and also inoculate with appropriate bacterial culture or biofertilizers (Reddy and Nagender,2013). While growing the pulses always use quality seed because such category of seed give vigorous crop which use biofertilizer and nutrients from the soil better than the poor crop seed (Chandra et al; 2013).

Realizing the importance of pulses by the Government of India in our dietary requirement and the agricultural production system, pulses development programme was brought within the ambit of Technology Mission on Oilseeds in the year 1990 (Tomar, 2008).

Gram:Gram occupies largest proportion of area under pulses and is mainly grown as rabi crop on conserved moisture. In some areas where the rainfall is low,the crop is grown with irrigation and more yield is obtained.It is a soil restorative crop and does not normally receive any moisture and fertilizer (Table 1).

Black gram: Black gram generally grown without irrigation during the rainy season either as a pure crop or in mixture with other crops.In some of the states,where winter is mild and somewhat warm, this crop is grown in rabi season on residual moisture or with light irrigation.Black gram can be grown on a variety of soils but relatively heavy soils. The crop is either un manured or inadequately manured with 7.5-12.5 tonnes of well decomposed farmyard manure per hectare. Well decomposed farmyard manure is always amorphous in nature and brown or black in colour (Table 1).

Green gram: Green gram is grown under more or less similar conditions as black gram. As is true to other pulse crops, manuring of green gram is also not done. In Peninsular India,the usual practice is to apply 5.0 to 12.5 tonnes of farmyard manure or compost per hectare as basal dressing in case of uplands. In lowlands generally no manuring is practiced. The crop very well responds to limiting on acidic soils. Careful attention is required to be paid to drainage in case of crop grown during the rainy season. Amount of atmospheric nitrogen fixation by gram varies from 85-110 Kg /hectare/year whereas in green gram and black gram ranges from 55-55 Kg and 50-75 Kg/hectare/year respectively (Table 1 & 2).

Table 1: Characteristics of effective and non-effective nodules

S.No	Character	Effective nodules	Ineffective nodules
1.	Colour of the nodules	Pink in colour or brown in colour with reddish pink inside	Usually whitish in colour or green
2.	What imparts brown or pink colour	Due to the presence of Leghaemoglobin	Lacking of Leghaemoglobin
3.	Characteristics of Leghaemoglobin	Properties similar to haemoglobin which is found in human blood	Problems like sodicidity, acidity, low amount of phosphorus, calcium, magnesium, molybdenum are responsible for lacking leghaemoglobin
	Fixation of atmospheric nitrogen	Capable of fixing atmospheric nitrogen	Not capable of fixing atmospheric nitrogen

Source: From R.D. Gupta, Farmers Forum, May, 2007.

Table 2: Range of nitrogen fixation of different pulse crops

S.No.	Name of the Pulse crop	Range of Nitrogen fixation(Kg/Ha/Year)
1	Gram(*Cicer arietinum)*	85-110
2	Green gram(*Phaseolus mungo)*	50-55
3	Soybean(*Glycine hispida*)	40-256
4	Red gram(*Cajanus cajan*)	168-200
5	Peas(*Pisum sativum*)	51-132
6	Lentil(*Lens esculenta*)	90-190
7	Black gram(*Phaseolusradiatus*	50-75

Source: Range values shown in the above Table were worked out by the author after consulting number of books, Journals of Research and Extension.

References

Chandra, Punetha, S and Kumar Chandan.2013. Quality seed: Characteristics and importance in crop production. Indian Farmers Digest.46(11):8

Gupta, R.D. 2007. Farmers should utilize Rhizobium inoculants to enhance legumes production. Farmer's Forum. 7(5): 24-27)

Gupta, R.D., Kaistha, B.P. and Kher D., 2007. Ecology of Azotobacter in soils of Himachal Pradesh and effects of its isolates on maize crop. Environment Ecology. 25S (4A):1304-1308.

Gupta, R.D. 2013. Farm women and future strategies. The Daily Excelsior, 49(361):6.

Gupta, R.D. and Sharma, A.K. 2004. Revive Green Manuring practice for higher food production. The Kashmir Times. 6(290):7

Reddy,G.K and Nagendra,T.2013. How to increase the nitrogen use efficiency. The Indian Farmers Digest.46 (11):22.

Sabharwal, Gourav. 2022. Soil health – Key to sustainability. The State Times 2(310):6.

11

Organic Farming and Land Degradation

Introduction

The Indian literature viz; Vedas and other related books show that ancient Indians used to impart great significance to the gardens and parks. Indian rulers especially, the kings, the princess vis-a-vis affluent people took special interest in making vegetable gardens in the houses and fruit tree gardens in open areas. In those days, it was considered to be the foremost duty of a king to maintain the beautiful gardens in the city and around his palace. It is even said that only he should be revered as an ideal king whose abode is provided with spacious gardens having full of fruit bearing trees (Krishna murthy 2012). As a matter of fact, the gardens also contained pools and tanks adorned with beautiful lotus blossoms, having a number of plants and bushes laden with large number of fragrant flowers etc. Kautilya has included gardens and fields under the term of "Vastu". So Vastu Sastra or the science of habitation may be taken to stand for laying out of gardens etc; including both horticulture and agriculture.

Merits of Planting Trees in the Gardens

1) Sacred puranas like Matsya and Agni purana bring out the merits acquired by planting of trees, by laying out gardens in good soils (Prasad, 2015) and sinking wells. In fact, they are treated as acts and public utility. They are as such considered and are regarded as actions of piety (Purta). Our fore fathers believed that trees are always resorted by Devas, Damavas, Gandharvas, Kinnaras Uragas etc, and one who lays out a garden or plant trees definitely gets the blessings of Lord Shiva, Vishnu and Brahma.

2) The Goddess Lakshmi always dwells for number of generations in the house of such persons who plant trees like Bael or Bilva (*Aegle marmelos*) Amla or Aonla (*Emblica officinalis* Gaertn Syn. *Phylanthus emblica*), Bahera (*Terminalia bellirica,*). Through several writings our ancestors have brought out the necessity of maintaining gardens and planting of trees of various kinds including fruit trees. In Sanskrit literature a number of statements are given which make people to realise the importance of maintaining greenery everywhere. For example, it is said "Dasakupasama Vapi Pulrah Dasa Putrasama Drums". "One tank

is beneficial to man like ten wells, one lake does good like ten tanks, ten lakes are beneficial like one son, but one tree is equal to having ten sons."

3) Puranas and several other poetic works are replete with reference to different types of fruits and flowers. In ancient India, as exists today also, flowers were given the top most place in the art of aesthetic make up of men and women. A number of varieties of colourful flowers are found in our country having lot of fragrance. Such kinds of varieties mainly consist of Mallik, Kelki, Padma, Sevantika, Champaka etc. This as a result, encouraged the manufacture of different types of perfumes, cosmetics.

"**Upavanavinoda**" is a well-known to realise in Sanskrit which deals with several scientific aspects of horticulture. Puranas like Agni and Matsya as also several other ancient works like, " Brhat Samhita," " Mansa Sallasa" furnish with lot of interesting information on this subject.

Various Aspects of Horticulture Science

Almost all aspects of horticultural science such as fitness of soil, sowing the seeds at proper time / planting of saplings, watering the plants and trees, application of manuring after preparing them in the pits beautifying the gardens and many other related topics are mentioned/ discussed at length in these treatises of ancient and medieval India and the techniques serve as guidelines even today.

Botanical Classification of Plants/Trees

The ancient works as stated above, to begin with, provide botanical classification of plants as vanaspati (which yield fruits without flowers). Lata those creeps on trees or on ground Gulma (shrub). Another classification is Bijaka (grown from seeds), Khandaja (that produces from cuttings) and Kanda **summudbhava** (grown from bulbs).

If we scan the Vedic literature, we find the mention of several morphological details related to plants thus revealing the fact that plant life was closely observed by the ancients scientifically distinguishing the various structure like mula, skanda, kanda, parma, pushpa and so on.

Importance of Soil for Garden

Vedic Indians have given much significances to the quality of soil/ ground prior to laying out a garden. Ancient scientists have stated three types of ground namely **Jangala** (dry), **Anupa** (wet or marshy) and **Samanya** (ordinary).

Each of these is again of many types depending upon the colour and quality. However, well levelled grounds (soils) are the best with a few exceptions. Apart from the above, in ancient India productivity of different types of soils was taken into consideration before planting trees. Works on horticulture provide a long list of trees classified on the quality of ground, best suited for their growth (Lal, 2021).

Nourishing of Plants

As nourishment of plants is essential to get more of flowers and fruits, so horticulturists knew different techniques of preparing special type of manure which would nourish the trees to yield lot of good fruits. Different manures fertilizers are prescribed for different plants. Details of preparing nutrient solution, treatment of plants in diseases etc; in works like "Vrikshayurveds" of Surapala, "A special type of fertilizing solution called, "Kunapajola," stands mentioned in other treaties.

The nourishing tonic for the plants was advised for supplying requisite nitrogen compounds, phosphates, potash etc. In Sanskrit poetic works we often came across with the word, "dohada, "which when applied in plants must be taken in the sense of special treatment manure.

Collection of Seeds and their Treatment

Several methods of collection of seeds and their treatment are described in ancient texts. Special care was taken for the treatment of different types of seeds to make them sprout out very fast. The rules to be followed by the persons (farmers, gardeners, orchardists) who saw the seeds or plants/ trees have also been specified. There was a strong belief that purity of heart and body, while sowing seeds should be maintained because plants, though they appear inanimate respond to the sentiments of the Gardner.

Planting or Sowing Time / Method of Planting

In the Vedas our forefathers have also mentioned auspicious asterisms and favourable months and days for sowing seeds or planting various types of saplings.

To make the gardens and parks more beautiful, ancient Indians planted trees in rows (avenues), in groups or clusters spacing them in such a way that the distance between two trees would be not less than ten feet (3 meters) and in between the trees flower plants were made to grow.

Protection of Plants

All plants and trees though they need good sunlight and open air deserve protection from dew, hail storms, smoke fire and heavy rains. For example, the flowering tree should be planted in the middle of the row of trees, fruit giving trees should be planted and protected with walls having ditches around them.

To protect plants/ trees from hailstorms they should be smeared on all sides with the ash of a tree hit by lightning. Ancient Indians looked after the plants and trees as members of their family. (As we can see in Kalidasa's portrayal of Shakuntala's love towards plants particularly the jasmine creeper vanajyotsna). Destruction of plants was regarded as serious offence and culprit was severely punished.

Insect Pests and Their Control

In Atharvaveda and in some Buddhist texts fungi and pestiferous insects are referred to and the gardner was punished for negligence. Even religious methods are mentioned in "Upavana Vinoda" for the protection of plants life from insects and moths etc.

Watering or Irrigating of the Crops

Newly planted trees should be watered both in the morning and evening. In sisira or hemanta trees, the watering should be done on alternate days, in spring every day and during summer both morning and evening.

Removal of Weeds

Any weed, creepers grass etc should be removed prior to sowing or growing next plants or trees. Removal of weeds should be done by hand weeding.

Many tips for the healthy growth of the trees are described in texts of Vedas. During rainy and autumn seasons when it does not rain a circular ditch (alavala) should be filled with water; sometimes the water poured will not get absorbed in the soil, it indicates indigestion and in such a case water should not be poured.

In dry places where rainfall was scarce it was advised to draw water from underground for irrigation. Since undergoing water had to be ascertained by several means like vegetative growth on the ground it developed into an art in ancient India. This art was called, "Dakargala," the term evidently refers to the determination of subsoil water (daka) with the help of wooden stick (argali), an art still practised in our country. From such information, from other works like Krishiparasara we can gather that ancient Indians were probably the greatest water harvesters in the world. They evolved a vast variety of water harvesting systems for horticulture, agriculture and even house hold purposes.

These practices speak of highly specialised surface hydrology and water management in ancient India.

Ascertaining of Underground Water

Presence of underground water was ascertained by the experts from the growth of certain trees planted in dryland regions. It was from the position of anthills near the tree's presence of living creatures under the tree, close growth of two or three different types of trees, growth of grass from the bent branch of a tree d so on. Works like Sivatattva Ratnakara have provided lengthy chapters on this particular topic.

Diseases of Plants / Trees

Ancient India had established that plants were prone to diseases like human beings and preventive curative medicines and treatments were prescribed to control them. Mention is made of diseases of trees born of three humours Vata, Pitta and Kapha and remedies are also prescribed when trees are eaten by insects, scorched by fire, broken by wind or hit by thunder bold.

In modern times scientists claim to have made astounding contribution to the science of botanical marvels by bringing into existence such as botanical products out of the combination of several things. A few treatises contain details though they appear to be unscientific which prove the fact that such botanical marvels were created in ancient India.

Creating botanical marvels by adopting different techniques is an interesting subject included in ancient works on horticulture and this has received sufficient attention in our ancient literature particularly in describing the beauty of heroines (Ashoka tree bearing flowers when kicked by beautiful Malavika).

A few books dealing with ancient Indian horticulture have included a section called, "Dohadantaram (Different manures / fertilizers) which mention a few other techniques to create botanical marvels and which may be called poetic conversions like kicking, spitting, embracing, gazing at dry trees by beautiful maidens to make them yield lots of flowers and fruits.

Ancient Texts on Horticulture

Ancient texts on horticulture contain a few sections bringing out different methods of beautifying gardens and parks. Creeper bowers, small lotus ponds, artificial caves, one or low swings, mounds and artificial peaks etc, be made. Provision should be made for the free movement of peacocks and other birds. Mounds in pounds were made to attract different species of birds to maintain the ecosystem.

Anthology

The anthology of different topics related to plant life, horticulture etc presents in ancient Sanskrit texts throw light on the highly developed state of arboriculture in ancient India.

There is no doubt, that our ancestors were thinking of religion and philosophy but along with it they evinced great interest in subjects especially horticulture and were endowed with the capacity and enthusiasm to do analytical thinking and scientific experiments in those days when adequate resources were wanting. A study of the Vedas and ancient works like upavasana reveal that trees and plants were worshipped by our forefathers. They gave importance to both religious and emotional factors related to plant life to save them from all sorts of destruction.

Vedic Indians, thus had established that plants are also living beings like human beings with the difference that their consciousness is internal or dormant, they too have feelings of pleasure and pain, when they are teased by way of applying unnatural articles or artificial things. It is point to mention that use of well decomposed bulky organic manures, have a direct effect on the plant growth like any chemical fertilizer. Well decomposed organic manure refers to the following the characteristics

a) It is black brown or totally black in colour

b) It is always amorphous in nature i.e. powder like in appearance.

c) Application of a foresaid type of manure to the soil will not only improve its structure, but also promotes water holding capacity and cation exchange capacity. It is more so in case of loose sandy soils by acting as a binding material.

d) Apart from the above, application of such kind of manure will encourage multiplication of micro-organisms i.e., bacteria, actinomycetes also known as thread bacteria and fungi as well as protozoa and nematodes (single cell type), and also macro-organisms.

e) Such kind of manure produces humus in the soil that helps in partial neutralization of the highly alkaline soil and, thereby reducing its pH value.

   By adopting such kind of practice i.e., organic farming improves the quality of fruits as well as those of vegetables and agricultural crops.

f) Organic manure also generates catalytic and simulative actions in the chemical processes in the soil and enhance bacterial activity of the soil fertility.

g) Poor soils which ordinarily cannot produce normal fruit tree or a crop under favourable soil moisture conditions should be manured with well rotten farmyard manure at about 10 to 12.5 tonnes ha^{-1} (4 to 5 tonnes $acre^{-1}$).

h) Application of farm yard manure in stone fruit trees like peach, plum and appricot during the months of December to January @ of 40 to 60, 25 to 30 and 10 to 15kg per tree respectively has proved useful in increasing their yield under Uttarakhand climate conditions (Singh et al 2004.)

The ancient life was nature friendly and a perfect harmony with the nature. Every house used to have a kitchen garden locally known as parabag (Ambad et. al; 2012) which was a striking feature of the rural life. It was composed of fruit trees, vegetables, medicinal and aromatic plants as well as flowers to meet the daily requirements of the family. This system still exists in Kerala but elsewhere it has been eroded because of so many reasons viz urbanisation, high cost of land and poor resources.

The fragmentation of the land among families decreased the per capita holding of the land and farm size became small. However, an ancient heritage of kitchen gardening can be at rescue and act as remedy to overcome the situation. Planting of dryland fruit trees has been found to be the most suitable phenomenon.

Planting of dryland fruit trees such as *Aonla or Amla, Jamun, Custard apple, lemon, Karonda, Girna, Mulberry, Phalsa, Drumstick* was the main practice. These fruit trees are grown along with vines such as bitter gourd, bottle gourd, cucumber, watermelon which fetch a very good earning to support the family. The botanical names for growing aforesaid fruit trees are shown in Table 3.

According to a study by the department of environment and forests all fruits and vegetables tested in six centres across the country had heavy metals not only in traces but in measured quantities and a significant percentage always exceed the safety standards (Pandit 2007). As organic farming contributes in improving the quality of the fruits and vegetables which have potential export market. Hence, all the major components amenable for organic farming are required to be adopted. Such components consist of addition of organic manure, following of crop rotation and enhancement of soil fertility through biological nitrogen fixation wherever it becomes possible. Apart from the above, crop residues can be utilized to make compost/vermicompost and farm yard manure. Incorporation of green leaf manure helps to convert crop residues into organic matter into the soil. Organic matter acts as a reservoir of plant nutrients and also helps in altering physio-chemical properties of the soils. As a matter of fact, the organic matter on fully decomposition produces humus in the soil, which as a result not only increases its water holding capacity and cation exchange capacity but also ameliorates its structure.

Table 3: Main organic fruit trees which can be grown in dryland areas

S.No	Name of the fruit trees	Botanical name of the fruit tree
1	Aonla or Amla	*Phyllannthus emblica Gaertn*
2	Jamun	*Syzygium cumini L*
3	Phalsa (Falsa)	*Grewia asiatica DC*
4	Karonda	*Carisa corandus* L
5	Girna	*Carisa spinarum*
6	Custard apple	*Annona reticulata L*
7	Drum stick	*Moringa oleifera Lam*
8	Mulbery	*Morus alba* L

References

Ambad, S.N Jagdand S.M and Kadam, J.H. 2012. Sustainable livelihood for rural family through organic fruit production in kitchen garden and dairy. Agrobios Newsletter. 10(9):62

Gupta, RD, Gupta Sushil Kumar and Bhardwaj. SD. 2016. Agrotechniques and uses of medicinal plants. Associated publishing company a division of Astral International Pvt Ltd New Delhi 110002

Krishnamurthy R 2012. Horticulture in ancient India. Bhartiya Vidya Bhavans 58(21): 1928

Lal, Banarsi 2021. Rural development through horticulture. The State Times 26 (201):6

Lal, Banarsi 2022. Promoting organic farming in Jammu and Kashmir. The State Times 27 (90):6

Pandit Jyotsna 2007. Organic foods The Daily Excelsior 43(295):6.

Prasad J. 2015. Healthy Soils for Healthy life. Indian Soil Survey Land Use planning (ISSLUP) News Letter 4(2):1.

Singh, R.K Verma S.K, Singh, H. Bhardwaj, P.N Regar R and Arya, R.R. 2004. Ways to boost the cultivation of stone fruits in Uttaranchal (Uttarakhand). Indian Farmers Digest Vol 37. No 5-6 pp 5-8 and 21.

12

Protecting the Degraded Land of the Indian Himalayan Region Through Organic Farming

Introduction

Since 2001, India's human beings population has crossed 1000 million, which is presently 1361 million and will became more than 1462 million during 2025. Such a large population of human beings can be fed and clothed only if we make the best use of the country's land and water resources. While water is replenishable natural resource. Contrary to this, the land is a non-renewable and irreplaceable resource. It, therefore, becomes an imperative to think of about the main problems of protecting/preserving and managing the land resources first of all and thereafter managing the water resources. Such an approach will, infant, prove a solution of both of these resources. It is because if the land of the country is looked after properly, the problem of water management would be automatically tackled (Singh, 2007)

Out of the total land area of India, about 329 million hectares (m ha) as many as175 m ha suffer from degradation, caused for the most part by the soil erosion, water logging, soil salinity and sodicity as well as acidity. This denotes that on an average at least two out of every 3 acres (1.2 ha) of land is in poor condition. It is also known that at least half of the sick land i.e., one third of the total is almost completely unproductive. Another one third is partially productive and it is only the remaining one third is in good health.

Similarly, now a day about half of the area in the Indian Himalayas has turned into a degraded condition (Sharma, 2004). This is more so in case of certain North Eastern Himalayan states as well as in those of North Western Himalayan states. Hence, it becomes obvious the even to superficial observer, there is no early chance of eradicating poverty even on surviving as a self- respecting nation so long as such a state of affairs of Indian Himalayan's land persists.

In light of the above, the main focus of this study relates to the various causes responsible for land degradation in the Indian Himalayan region, methods for control it, and suggestions and priorities for the future.

Causes for Land Degradation

There are a number of causes responsible for land degradation. Some of them which are very important are presented below:

Deforestation and Soil Erosion: The most serious threat to the soil/land degradation is posed by deforestation. The deforestation in the central India has deprived of tribal people of their conventional sources of food. The deforestation of the Himalayas through commercial trees felling have created serious problems for the country. As a result of deforestation of Jammu and Kashmir Himalayas, our waterfalls, springs have already been dried up. Deforestation spells cultural death for the estimated 140 million tribal people who as hunters-gatherers or agriculturalists or rubber tappers depend entirely upon forests for their sustenance. Millions of them even now are being forcibly displaced and resettled, generally under Government's sponsored schemes.

Deforestation resulted in lot of soil erosion. Soil erosion is widespread and desert like conditions in the Himalayas including Jammu and Kashmir. Kashmir valley is the worst affected. Not only this, soil erosion from the bare hills is silting the rivers in fury and frequency.

It is worthwhile to mention that soil erosion has already affected over 150 million hectares (M ha) of land. Over half of this is under cultivation (Singh, 2007). In India although about 75 M ha are described as forest land, 40Mha are without good tree cover. Between 1950 to 1980, the unregulated diversion of forest land to non-forestry purposes, has resulted into a loss of about 4.5Mha of forest land. On an average it works to be an annual diversion at the rate of 1.5 lakh ha. Since the forest conservation came to fore in 1980, the annual diversion has brought down to about 25000ha (Anonymous, 2000).

It is impossible to determine the loss caused to the economy by these processes. It is for, apart from the progressive loss of productivity which results from the soil erosion, the displacement of top soil causes the premature siltation of costly and often irreplaceable reservoirs to the detriment of irrigation, flood control and hydel generation. Again, displaced soil raises the beds of the rivers, streams thus reducing their water carrying capacity. This in turn creates more floods.

Various Types of Soil Erosion

The soils occurring in the hill regions of the southernmost range of the Himalaya known as Siwalik (Shivaliks) pose a severe problem of: i) Sheet erosion and ii) Rill erosion.

i. *Sheet erosion*: Sheet erosion may be defined as more or less uniform removal of soil in the form of a thin layer or sheet like in appearance by the running water from the entire land surface of somewhat large area. It is particularly on gently sloping land (Gupta, 2009). This type of soil erosion is causing a great damage to submountain or low hills of Siwaliks of Uttarakhand, Himachal Pradesh, Punjab, Haryana and Jammu region of Jammu and Kashmir state (Union Territory)

It is point to mention that the top soil has been removed from entire area of about 4,89,266 ha engulfing a number of areas of Jammu submountain Siwaliks locally known as Kandi belt on Mansar to Suriansar, Suriansar to Sidhra, Samba to Mansar Dhansal to Jindhra and Bhalwal to Amb Garota (Gupta, 2002; Gupta 2005).

ii. *Rill erosion*: Rill erosion is also known as Micro-Channel erosion. When sheet erosion goes on unchecked, the runoff water, laden with soil particles run into small channels, spreading out quite akin to fingers like structures over the fields, cause rill erosion.

Like sheet erosion, the soils of sub-mountainous Siwaliks also pose a severe rill erosion problem due to dip slope and escarpment topography. This type of erosion can be seen in a number of broad longitudinal valleys known as "Duns". Dehradun of Uttarakhand Himalayas is situated in one such valley. In Jammu's Siwalik region of Jammu and Kashmir. Himalayas, such valleys are in Udhampur, Jhajar Kotli and Dhansal areas, which are locally called "Dadhals". Owing to high intensity of rain storms these soils are annually lost about 106 tonnes(t) ha^{-1} year, even in slope of about 2 to 3 percent in clay loam soils, experiencing annual rainfall of 1400-1500mm-in the absence of soil water conservation measures (Gupta and Banerjee, 1991) or even more in soils of Karewas of Kashmir (Gupta, 2005), where the figure of loss of soil are confined to locally called as, reported to loss found up to 225 t ha^{-1}.

iii. *Gully erosion:* When many thin channels grow into wide and deep channels are called as gullies. These not only eat away the farm lands but also cause erosion in the woodlands.

The areas where both surface and subsurface soils are fragile and easily cut by the flowing water, gullies with vertical walls are developed. On the other hand, if the subsoil is resistant to rapid cutting because of its toughness or heavy texture, especially where underlying geological materials is not soften than upper horizon, gullies develop on sloping banks and assume a "V" shaped gully formation. It is remarkable to note that many of the soil series of Siwaliks lying in the Jammu, Kathua and Udhampur districts of Jammu Province, Jammu and

Kashmir state, are suffered from gully erosion with frequent occurrence of "V" shaped gullies as reported by Gupta (2009).

The cold arid region of India, consisting of Ladakh Union Territory, Lahaul and Spiti as well as Kinnaur of Himachal Pradesh, also known as Trans Himalayan Region of the country, is characterised by the gully erosion. The rill erosion is also present especially in Lahaul spiti area due to snow melt accumulation in small channels over the sloping lands (Gupta et al., 2011). Side valley or Ladakh carved out by important tributaries of the river Jhelum the Sindh, the Lidder, the Vishav, the Rmbira, and the lolab, has undulating slopes which has much affected with gully erosion.

iv. *Glacial erosion*: The glacial erosion mostly occurs by plucking, grooving, scratching and chiseling by the snow. This type of erosion is predominant in cold regions of the country including cold arid Union Territory of Ladakh, where mean annual temperature is found below zero degree Celsius. As such glacial erosion has been mostly found in higher Himalayan region, covering the states of Uttarakhand, Himachal Pradesh and Ladakh Union Territory i.e., North West Himalayas as well as the Trans Himalayan Region (Kinnaur and Lahaul spiti of Himachal Pradesh.

The glacial erosion is quite predominant in cold arid region of the country. It is because of large number of peaks of this region remain under perennial snow or receive snow for most of the time during winter months. Whenever there is movement of large mass of ice down slope it brings with huge debris causing lot of soil erosion. It is characterized by furrowing, cutting, ploughing and scouring, which are referred to as "exaration".

Grinding and rubbing also called as "Glacial abrasions" or "detersion". "Detraction", is the last type of glacial erosion, which is a typical form of disturbance resorted by glacial erosion (Gupta, 2009).

Glacial erosion is prominent in the aforesaid areas. A large number of peaks are under permanents snow cover and as the glaciers move, process like plucking and nivation, glacial abrasion start operating which in turn accelerate soil erosion. During the summer months, the glaciers melt and the rills, streams and rivers swell and increase flash floods. The flash floods inflict a lot of damage to beds of the rills and gullies. The total area under the glacial erosion in Ladakh Union Territory and other areas of Jammu and Kashmir has been reported to be 4455Km^2.

A large-scale fluvial erosion further supplements glacial erosion. The big pinnacles on the banks of the river Indus exhibit the story that how relentlessly the surrounding lands have been eroded and found their entry into the river direct.

Reduction of Recharging of Underground Water

Apart from soil erosion, the excessive runoff of rainwater along denuded slopes reduces the recharging of the ground water aquifers. Thus, an invaluable resource which would otherwise have been available for round the year use as ground water is lost to the sea. Often this runoff sometimes prior to reaching sea wreak terrible flood havoc in many of the areas of the country. Floods are thus, caused by poor land management. The occurrence of floods has now become a serious problem in a number of states of the country not only due to excessive runoff loss during rainfall but also because of change of climate. As for instance the rainfall between August 9 and 15, 2018, was over three times higher than central forecasting agency, wrought a massive flood in Kerala. In these floods 483 persons were died and lakhs of people were displaced. Occurrence of floods in Uttarakhand and Kashmir valley of J and K state during the years 1913 and 1914 are the other examples in this connection.

Loss of Soil Through Soil Erosion

As has been stated earlier that various kinds of soil erosion have been recognised depending upon the predominant eroding agencies by which the soil erosion is caused. They mainly consist of splash erosion, sheet erosion, rill erosion, gully erosion and ravine formation, which stand appeared in literature (Gupta, 2002; Gupta et al. 2011). The soils of the north western Himalayan states (Jammu and Kashmir, Himachal Pradesh, Uttarakhand and submountain areas of Punjab and Haryana are badly affected with all kinds of soil erosion.

In many hilly regions the loss of soil is as high as 80 tonnes ha^{-1} $year^{-1}$ or even more especially in Shivalik's (Sharma, 2004). In Jammu and other Shivalik's, loss of soil through water erosion has been found to be the extent of 168 tonnes ha^{-1} $year^{-1}$ (Jamwal and Gupta, 2007). Soil erosion in some of the catchments of Punjab, Haryana, Himachal Pradesh and Jammu region is as high as 225 tonnes/ ha (Singh, 2007).

Wind Erosion in Trans Himalayan Region

Because of very low rain fall and cold arid climate in the Trans Himalayan region viz. Ladakh Union Territory and Lahaul Spiti and Kinnaur areas of Himachal Pradesh, the wind erosion is quite prevalent. Wind erosion not only robs of the richest soil of the land but sometimes planted crops are also blown away. Leaving behind exposed roots or these crops may be covered up by the drifting debris. Wind velocity greater than 30km hr^{-1}, (quite common in the Ladakh region) is responsible to blow sand from some cold arid areas to spread on the fertile lands, creating thereby reduction in the agricultural productivity.

Out of the total land 1.36 M ha are affected by the wind erosion, 0.78 Mha (57%) and 0.55Mha (40%) suffer from moderate and strong erosion respectively. These areas are mainly confined to the eastern aspect of Ladakh Union Territory.

Desert Development Agency, Leh Ladakh has identified a number of decertified areas where sand dunes are of common occurrence. These sand dunes are stabilized by planting shrubs like *Hippophae* spp and *Meicaria* spp. These plants can survive even on a limited moisture. *Seabuck* thorn (*Hippophae* spp) has a unique quality to with stand and adverse temperature upto 60° Celsius below freezing point. High velocity winds and a rainfall of 300mm cannot harm the growth of this tree.

Impact of Wind Erosion Loss of Soil

Annual losses higher than 300 tonnes/acre (750 tonnes/ha) have been estimated for highly erodible bare sandy soil (Sharma, 2004). An entire furrow slice (about 25 tonnes/ha) could he blown away in three or four years at this rate if soil was removed uniformly from the entire surface.

Loss of Plant Nutrients

As colloidal clay and organic matter are the main components of plant nutrients both primary and secondary, i.e., N, P, K Ca, Mg, S and Fe, Cu, Mn, Zn, Mo, B and Cl respectively. Hence, when colloidal clay and organic matter are lost in dust storms, lot of plant nutrients present in them are also lost. Thus, the soil fertility is lessened.

Loss of Productivity

As the soils become less fertile due to loss of soil fertility so crop productivity will also become low. It is because soil fertility, good management practices availability of water supply and a suitable climate contribute soil productivity. Soil fertility denotes the status of plant nutrients in the soil, while soil productivity connotes the resultant in various factors influencing crop production, both within and beyond the soil.

Effect of Soil Erosion on Agriculture

Unabated deforestation and overgrazing of the pastures (grasslands) in the Himalayan regions of India including Siwaliks have increased a considerable level of soil erosion. This as a result has posed a major threat to the agricultural production as well as to the hydel power of project and normal climate of the region. It is remarkable to mention that soil erosion problem is not only being faced in the Himalayan tract of the country but all over the world today.

This process is seriously not only damaging the food grains productivity but also posing a problem in power generation in dams and the irrigation projects besides harming the productive wetlands of the country (Gupta, 2016).

Loss of Top Soil

The loss of top soil (0-15cm or 22.5cm) which is extremely essential for growing plants/trees, is directly affecting the green cover on the land and agricultural production. The rate and speed of soil loss every year has been found several times more than the soil formation and if this process is allowed to continue, the human life on the earth would be hit badly owing to the following reasons:

i. Low availability of food grains.

ii. Less available of fruits/vegetables.

iii. Low oxygen supply as felling of trees is going on continuously especially in the Siwaliks of Jammu.

It is point to mention that once rich green Siwaliks of Jammu before 1947 have now become mountain deserts because the people and the Government have not taken note of the ecological problems of the Himalayas including Siwaliks (Jamwal and Gupta 2007). As for example, up to 1960 Phalsa (*Grewia optiva*) used to grow abundantly along with other fruits like girna *(Carissa spinarum*) Karonda (*Carissa carandus*), amla (*Phyllanthus emblica*), mango (*Mangifera Indica*), ber (*Zizyhus jujuba*), loquat (*Erybolrio japonica*), and Jamun (*Syzygium cumini*) as reported by Gupta and Banerjee (1991) and Gupta et.al;(2016). The above said fruit trees used to grow commonly along with various dwelling species of wildlife, particularly in the Kandi belt of Jammu (Gupta, 2016). The author himself recollects those days when, Phalsa selling halkers, used to speak "*Aa Gaya Phalsae Wala, Thund Panda Je Phalsa, Garmi Goanda Je Phalsa*"।

Apart from growing of fruit trees as stated above, the Kandi belt of Jammu was also popular for growing of a number of medicinal plants like Harar (*Terminalia chebula*), Bahera (*Terminalia balerica*) besides Amla (*Phyllanthus emblica*). Arjun (*Terminalia arjuna*) tree was another medicinal plant which used to grow on large scale and still grow in submountainous areas of Siwaliks of Jammu, Punjab, Haryana, Himachal Pradesh and Uttarakhand, Neem

(*Azadirachta Indica*), Ashwagandha (*Withania somenifera*) and Rattanjot or Jangli arandi or Kala Aranda (*Jatropha curcas*) are the other medicinal plants which used to grow in Kandi belt of Jammu, Kathua, Udhampur and Rajouri on large scale. This stand proved by growing the aforesaid medicinal plants in herbal gardens set up through a centrally sponsored scheme, "National watershed development projects" (NWDPRA) was started in 1990-91. Division of Ministry of Agriculture was involved in monitoring this project aided and funded by World Bank Danish Development Agency, United Nation Development Programme and Council for Advancement of People's Action and Rural.

Green methodology: Green or so to say afforestation of all the waste lands is the paramount need of the hour and covering all such lands with suitable vegetative species to prevent the erosion, water conservation, removal of salinity from the soil as well as to restore soil micro biological activity (Khajuria, 2022). Apart from this, it will help to lower the soil temperature, increase the water percolation and increase the agricultural productivity. Duneuded hills slope in the Himalayan Shivalik's have their own sad stories to tell as in such areas, most of the top soil has been badly eroded to the extent of its total washaway which hinders planting of suitable tree species on such barren and denuded areas. So, treating such areas, introducing of shrubs is the first step to go ahead and every precaution is to be ensured to avert biotic pressure.The shrubs must be preferably of atmospheric nitrogen fixing ones and of the few are: Bana (*Vitex negundo*) Branker (*Adhatoda vasica*), Madhuaeshthi (*Glycyrrhiza glabra*). After the soil reclaimed, the most suitable plants are required to be planted. Preference must be given to grow local tree species which are mentioned above.

Technology (CAPART), Government of India. The programme was conducted under the supervision of Major General Goverdhan Singh Jamwal (Retired) through Paryavaran (Environment) Sanstha (Jamwal and Gupta, 2007). Under this project growing of phalsa was also tried in village Suchani, Samba District which was proved as a "Mantra" for the Shivalik's like the Leh Berry (*Seabuck* thorn) of Ladakh, based on this success story, Indian Institute of Integrative Medicine (IIIM) has already installed plant and machinery for processing Phalsa products as well as those of Amla and Aloe vera.

Soil and Water Conservation Measures

A large portion of arable land of the North West Himalayas especially of Kandi belt of Jammu Province as well as that of Kashmir Karewas(also known as kandi), Punjab,Haryana as well as Himachal Pradesh of Kandi belt has become unfertile so the below mentioned soil and water conservation measures or practices are required to be adopted to enhance the fertility of the soils and thereby, their productivity.

1) *Contour Farming:* As ploughing up and down the slope land courses quick run off of water and loss of surface soil by erosion, whereas ploughing, sowing/planting, harvesting and other operations done along the contours always check the flow of water straight down the hills which thereby, keep controlling the soil erosion. Experiments conducted in this line have shown that contour cultivation alone, reduces run off from 40 to 54 per cent and soil loss from 30 to 20 tonnes per hectare as reported by Gupta and Banerjee (1991). Contour farming is also required to follow, wherever it is possible in the eroded catchment area present in the vicinity of Wular Lake to prevent degrading situation of the land which is one of the main causes of siltation in the Wular Lake (Gupta, 2016).

2) *Terracing*: It is another method to prevent the run off of water. It consists of dividing the land slope into a series of small flat fields by means of terraces which are placed in such a manner that they can catch and hold water. This checks the run off and allows the water to percolate deep. This breaking up of long slopes into small mini-watersheds channelises water to the sides of the fields and as such soil erosion is controlled. Generally, terraces have been classified into the: 1) Broad channel type,

3) Ridge type and 3) Bench-type. Bench tracing is resorted where the slope is more than 12 per cent whereas Broad Channel type and Ridge type terraces are required for those areas which have mild slope. There is ample evidence that bench terracing treatment can considerably reduce run-off and soil loss occurred due to erosion than any other conservation methods. Gupta and Banerjee (1991) reported that bench terracing can minimise runoff 50 per cent and soil loss 98 per cent as compared to other kinds of terraces.

4) *Strip cropping*: Strip cropping is another suitable method applied to control soil erosion. Strip cropping consists of growing of different crops in alternating strips such that they serve as vegetative covers or barriers to prevent soil erosion. There are four types of strip cropping namely: a)

Contour strip cropping, b) Field strip cropping c) Buffer strip cropping and d) Wind strip cropping. These are described in details as under:

i) *Contour Strip Cropping*: It consists of growing of erosion resistant field crops on slopy lands, in regular long and narrow strips of variable widths. The strips are placed across the slope on the contours.

ii) *Field Strip Cropping:* In this case crops are laid out in strips across the general slope but not exactly following the contour. This is generally adopted when the soil has high permeability and the slopes are uniform or the land is undulating without well- defined slopes. It has proved quite effective on regular slopes. But on irregular slopes, field strip cropping has not proved useful.

iii) *Wind Strip Cropping:* This has proved beneficial for controlling wind erosion. In this system, the strips of crops are laid out at right angle to the direction of the prevailing winds.

iv) *Buffer Strip Cropping:* These are strips of permanent crops grown in the soils and are usually located on steep and badly eroded areas which do not fit into regular crop rotation.

It is remarkable to note that out of these four kinds of "strip cropping" contour strip is found very effective and a practical means of protecting land from soil erosion on well drained, soils having slopes even up to 16 per cent. This kind of strip cropping has been successfully used on lands having capability classes like II e, III e and IV e.

Contour strip cropping is adopted to well drained cultivated soils on sloping land, where hard beating rains are frequent and soil erosion is a hazard. However, in high rainfall areas contour strip cropping with gradient of 1.2 per cent is more desirable to lead the water into a grassed waterway. Steepness of the slope, texture of soil and the normal amount and intensity of width of rains are important in determining the erosion resisting and erosion permitting crops. Steeper the slope, greater is the width of erosion resisting crops. In areas on slopes less than 6 per cent width of about 30m is adopted. For slopes 6 to 10 per cent the width is about 24m and for slopes 11 to 16 per cent it is about 15m.

4) *Mulching:* Mulching either consists of stirring the upper surface of the soil to a depth of about 2-3cm or spreading litter, plant leaves and crop residues on the soil with the objective to conserve the moisture and reduce its loss through evaporation as well as to protect the soil from blowing away for by wind and water erosion. Crop residues mostly consist of straw, cotton stalks, leaves, saw dust, pine needles, coir dust and other material like polythene films or certain special kinds of paper in the tree basins and interspaces between trees. The main advantages or benefits that occur through mulching are given below:

(1) *Protection of Soil:* The soil from blowing and washing away by wind and water is protected. A significant reduction in run off and soil loss can be achieved by using grass and leaf litter mulch 4 tonnes /ha on the sloppy lands of Dehra Dun (Verma et al. 1979).

(2) *Conservation of Soil Moisture:* In unirrigated areas the rainfall is boon. The soil moisture is advantageously utilized for sowing of crops. Under proper soil moisture conditions when soil mulch is done by light harrowing, the capillary connection of the top surface layer with the sublayers is broken which as a result checks the upward movement of the soil moisture. Thus, mulch is an important mean /source for conserving soil moisture.

(3) *Addition of Organic Matter:* Stubble mulches besides checking soil erosion also supplies organic matter (Gupta et al. 2011). Organic matter after decomposition produces humus, i.e., humic acid, fulvic acid and humin which is helpful in improving physio-chemical properties of the soils (Gupta et al. 1992). For example, soil structure is a result of the interaction between inorganic (clay minerals, and amorphous clays) and organic colloids (humic acid, fulvic acid, humin etc). The organic colloids are responsible for soil aggregate stability.

Impact of Mulch on Rainfall and Soil pH: The mulch layer will assist soil to soak in water slowly and it will also lessen the impact of rainfall as it penetrates through to the soil. The mulch layer will also stabilize the affected soil to be healthy and neutralized. (Raina and Chohan, 2011).

Mulch Planting and Weed Control

Mulch Planting has been found to be effective means of weeds control (Kumar et al. 2011).

Reduction of Irrigation: Frequency Besides reducing surface run off, preventing soil erosion, adding soil organic matter, mulching also reduces irrigation frequency for crops, fruit trees and vegetables production (Kumar et al. 2011). Prevention of The Contact of Fruits with Soil and Thereby Preventing the Soil Borne Diseases

As mulching prevents the contact of the fruits with the soil, thus preventing the soil borne diseases like fruit rot (Kumar and Rattan, 2011).

Crop Rotation

Crop rotations have an important place in scientific soil management and crop husbandry. In any cropping pattern designed for the maximum exploitation of soil and water resources for obtaining or getting higher yields, the crops are required to be rotated systematically.

What is Crop Rotation?

Crop rotation is an order or a sequence in which the chosen or selected cultivated crops follow one another in a set-cycle in the same field over a definite period for their growth. The period may vary, from one year to three years or even more. Rotation of crops are planned, carefully after considering the type of the soil, climate, irrigation, the nature of demand in the market for agricultural produce and price. The main advantages or benefits obtained from adopting proper crop rotations are the following

Advantages

i. The soil is managed properly with regard to its utilization and preparation for growing each crop. This aids in keeping soil texture and structure for proper conditions.

ii. The soil fertility is considerably replenished by including restorative leguminous crop in between the exhaustive cereal or fibre crops or by raising for burying a green manuring crop included in the rotation.

iii. The drain upon soil fertility is kept uniform by following a regular system of crop rotation,

iv. The possibility of some soils developing soil toxicity is removed, and healthier soil condition is maintained.

v. The soil microorganisms are able to play their full role in enriching the soil by their action on the soil organic matter.

vi. The available irrigation water facilities can be best utilized by following intensive crop rotation, or by following partly intensive crop rotations and or by including rotations of cash crops require more water for their growth or by getting catch crops.

vii. Economy is exercised in the use of irrigation water particularly where the cost of irrigation is high, by growing a crop of low water requirement after a crop requiring more water and thereby, utilizing the soil and subsoil moisture.

viii. Under dry farming conditions suitable crop-rotations permit sufficient time for conserving soil moisture and developing soil fertility necessary for the successful cultivation of crops.

ix. Crop rotations permit the cultivation of deep-rooted crops followed by shallow rooted, exhaustive crops followed by restorative and cereal crops followed by leguminous ones and thereby allow a balanced utilization of the soil fertility.

x. Crop rotations provide for an easy eradication of weeds during the period of preparatory tillage for the crops and reduce the damage to them.

xi. Crop rotations also help in destroying fungal diseases and insect pests to a considerable extent which attack and thrive up on some crops.

xii. Well planned crop rotations, render cultivation convenient and remunerative to a large extent.

xiii. Crop rotations are accompanied with the notation of fields falling under each set of crops according to seasons. This leads to convenience in the performance of agricultural operations and supervision.

xiv. Crop rotation and double cropping practice is specific to remote locations of Lahasaul (Miar nullah) in Himachal Pradesh. Rotation starts with barley in the first year, and buckwheat during the second. This helps to meet the challenges of limited availability of organic manure for the successful management of soil fertility levels as well as pest management.

Grouping of Crop Rotations

Crop rotations are mainly grouped into two classes :i) Rotations on irrigated lands and ii) Rotations on rainfed (Barren) lands.

Rotations on irrigated lands are further classified into: i) Rotations of medium intensity with percentage of crop intensity varying from 100 per cent to 180 percent ii) Rotations of high intensity with intensity of cultivation varying from 180 percent to 300 per cent or even more.

The intensity percentage can be worked out by multiplying the number of crops in a rotation with hundred and dividing by the number of years taken to raise these crops.

Main Crop Rotations Used to Be Followed Before Green Revolution: Medium intensity rotations recommended generally include most of the food and commercial crops which are described as under:

1. Green manure (Guara or Sanhemp) - Wheat- Maize-Senji-Sugarcane. This is a standard rotation for these crops, allowing to grow crops in

three years, green manure being excluded as it is buried in the soil to add its fertility. In this crop rotation intensity works out to be (1004)/3=133 per cent. The rotation also includes one restorative leguminous crop of Senji.

2. Green manure (Guara or Sanhemp) Wheat - Maize-Senji or Berseem - Cotton. Intensity percentage is 133. Senji or Berseem in this case is a restorative crop prior to growing cotton.
3. Green-manure (Guara or Sanhemp), Wheat - Maize-Gram - Cotton. Intensity percentage of this crop rotation is also 133. In this crop rotation, gram works as a restorative leguminous crop before growing cotton. This rotation of crops was very popular where water supply during Rabi season used to be less in quantity.
4. Green manure (Guara or Sanhemp) -Wheat - Chari Guara-Gram-Cotton-Senji- Sugarcane. Intensity percentage of this cropping pattern on crop rotation is 150. In this case, we get restorative crops, one before cotton and other before sugarcane, in addition to green manure before wheat crop.
5. Green manure (Guara or Sanhemp) - Wheat-Toria or Sarson-Cotton. Intensity percentage of this crop rotation is found to be 100.

   It is concluded from the afore said “Rotational Farming” that adoption of the leguminous crops as “Green Manuring” is very essential. It is because inclusion of leguminous crop in a crop rotation as green manure not only enhances the yield of the crops but is also found to increase fertility of the soils and help in reclamation of saline and sodic soils (Gupta et al. 2011 and Gupta and Arora, 2016).
6. Green manuring of Sanhemp and dhainche must be done during early part of the monsoon in soils of Kandi belt to check soil and nutrients losses (Gupta and Banerjee, 1991).

Organic Manuring: Constant or continuous use of soil for crop production renders it poor in plant nutrients, even if the cultivated crops are sown under well planned crop rotations. As a matter of fact, this continuous growing of food grain crops, fruit trees and vegetables drains upon the soil fertility, which leads thereby, their poor stands and low yields. Moreover, the plants become weak in vitality and less resistant to the attack of diseases. Therefore, then, arises the need for adding plant food to the soil from external sources to make good the loss. The external sources of plant nutrients mainly consist of organic manures and fertilizers. These days, however, use of organic manures is preferred due to popularity of "Organic Farming". Fertilizers are not used. The main advantages of the Organic Manures are as follows:

1. Organic manures improve the structure of the soil and also its air cum water permeability.
2. Organic manure promotes the water holding capacity of loose sandy loam soils by acting as a binding material, while they encourage granulation of the heavy soils, i.e., clayey soils and facilitate drainage and easy pulverisation.
3. Organic manures encourage the multiplication of the microorganisms like bacteria, actinomycetes and fungi. All these organisms help in decomposing of organic matter and, thereby, release plant nutrients for growth of plants.
4. Some bacteria like Rhizobia and Azotobacters fix atmospheric nitrogen through symbiotic and a symbiotic processes in the soils for the growth of the plants.
5. Microorganisms produce humus in the soils which helps in partial neutralization of the highly alkaline soils and thereby reducing their pH value.
6. Organic manures also generate catalytic and stimulative actions in the chemical processes in the soil and promote bacterial activity on the soil fertility and thereby its productivity.
7. In cold arid zone of Ladakh, the placement and method of application of organic manures is one of the management tool for reducing losses of plant nutrients of the manures. Organic manures in the fields of the aforesaid cold arid zone are generally disposed of to the fields in the form of small heaps which remain as such there for days together. It results into the much more losses both of primary (N, P and K), secondary (Ca, Mg, S) and micronutrients (Fe, Cu, Mn, Zn, Mo, B and Cl). It is therefore, advisable to scatter the manure uniformly in the fields instead of dumping in the form of heaps. As plant nutrients management strategy is a pilot project in itself which always demands multidisciplinary access to Agronomist, Soil Scientist, Plant Physiologist and Plant Breeder so farmers must consult frequently to these Scientists while preparing / applying organic manures in the fields.

Imparting Right training: If the farmers of Jammu and Kashmir are required to be imparted right trainings in the organic farming accompanied by right resources and marketing techniques, thus, they will be able to grow organic fruits, vegetables and food crops and sell them across India and abroad. Then organic farming in Jammu and Kashmir will surely have a bright future. To overcome the use of chemical fertilizers such as urea, diammonium phosphate and murate of potash,the experts are emphasizing to develop the culture of organic farming or organic agricultural practices as it has the potential to address environmental, health and sustainability issues. It is remarkable to note that organic/agriculture is not new to Jammu and Kashmir as it is practised on approximately 50 thousand hectares of land and erstwhile state had organic certified area of 22,316 hectares (Sabharwal, 2022)

Principle of organic farming: The key principles of organic farming primarily consist of sustainability (ecological, economical and social), traceability and natural productivity. The organic farming is also associated with support for principles beyond cultural practices such as fair trade and environmental stewardship, although this does not apply to all organic farms and farmers. Organic farming in Paddar(Kishtwar and Udhampur):Taking the lesson from the Paddar belt of the Kishtwar district,the farmers of Udhampur have also decided to start organic farming. Agriculture land in some parts of Udhampur has virtually become barren due to the excessive use of fertilizers and pesticides as reported by Mahotra (2009). The farmers of this belt has ultimately realized that without returning to the centuries old traditional system of farming, it is not possible to increase their crop production. It is pertinent to mention that the farmers of Paddar have been producing organic vegetables which have great demand in Jammu and Kashmir as well as in national market. According to Kumar (2022) the use of agrochemicals in the agriculture has posed severe threat to the agriculture and allied sectors, environment and its resources including human beings and animal health. As such, there is great need for an adoption of organic farming. Low usage to no usage of chemical fertilizers and pesticides will save our soils, water bodies and rivers from pollution vis-a- vis animal health and human beings. The lower Shivaliks, intermediate hills and upper reaches can supply variety of organically produced fruits, like walnuts, cherry, pears, apples produced in the Kashmir valley and Jammu region without use chemical fertilizers (Raina, 2021). There is a great potential of production. brand promotion and marketing of organically produced local specialties like Bhaderwah Rajmash, Marwah Rajmash, Kala zeera, saffron of Pampore and Kishtwar. In Reasi district Jammu Province of J&K, there are many potential farmers who are well acquainted with applying organic practices in their fields to produce the organic food. The scientists of KVK have imparted trainings to them in preparing Panchgavya, Beejamruth, Jeevamruth vermicompost etc. (Lal and Tandon, 2022). The central Government launched a scheme, Jammu and Kashmir Arogya Gram Yojna, initially, 1000 villages of Kathua, Jammu and Udhampur were planned to be covered under the scheme for growing of aromatic plants purely organically. The land for its cultivation was supposed to be identified by the Council for Scientific and Industrial Research (CSIR) scientists and Aroma Experts (Sabharwal, 2022) while launching the scheme, science and technology minister Jitendra Singh claimed that farmers would be able to earn one lakh to 1.5 lakh per annum per hectare through th programme. The central government vowed to initially spend over 25 crores on this scheme.

Conclusion

It is concluded from the organic manures description that organic manures furnish the soils with organic matter which in turn promote their soil aggregation and thereby, ceases soil erosions and runoff losses. Water and nutrients retention in the soils is enhanced. Organic manure should be applied during kharif season. The organic manures used must be well decomposed i.e., it must be in powder/amorphous form and black brown or dark brown in colour. It should always be prepared in the pits. Organic manures, i.e., farmyard manure compost, vermicompost should be applied in thin layers after harvesting of wheat crops during Kharif season which will act as mulch.

References

Anonymous - 2000. Wastelands or wasted lands. The Daily Excelsior. 36 (203):6.

Gupta, R.D., Gupta, J.P. and Singh, H. 1992. Problems of agriculture in Kashmir Himalayas (pp- 53-67). In conserving Indian Environment (Ed. S.K. Chadha) Pointer Publishers, Jajpur (Rajasthan), India.

Gupta, R.D., Gupta, S.K. and Bhardwaj, S.D. 2016. Agrotechniques of medicinal plants. Associated company, Division of Astral International Pvt. Ltd., New Delhi

Gupta, R. D., Kumar, A. and Sharma B.C. 2011- Management of Problematic Soils of Northwest Himalayas for Sustainable Hill Agriculture. In Sustainable Hill Agriculture - An overview (Eds. Anil Kumar, B-C. Sharma and Vikas Sharma), Published by Agrobios (India), Agro House Behind Nasrani Cinema Chopasani Road, Jodhpur, Rajasthan.

Gupta, R.D. 2002. Gaddis of Himachal Pradesh. The Daily Excelsior.38 (187):1.

Gupta, R.D. 2005. Environmental Degradation of Jammu and Kashmir Himalayas and Their control. Associated Publishing Company Karol Bagh, New Delhi - 110005 (India).

Gupta, R. D. 2009. Glimpses of North West Himalayas. Published and printed by Radha Krishan Anand & Co. Pacca Danga, Jammu Tawi-18 0001, J&K (India).

Gupta, R.D. 2016. Wildlife and Wetland. Ecosystems of Jammu and Kashmir Himalayas. Published and Printed by N.R. Books International Pacca Danga, Jammu Tawi -180001, J&K. (India, Phone-01.91.2546691).

Gupta, R. D. and Arora, Sanjay. 2016. Salt affected soils in Jammu and Kashmir. Their management for enhancing productivity. Journal of Soil and Water Conservation 15(3):192-204.

Gupta, R.D. and Banerjee, S.K. 1991. Problems and Management of Soils and Forest Resources of North West Himalayas. Mahajan Book Centre, Last Morh Gandhi Nagar Jammu, J&K (India), Now Sahil Book Depot, Last Morh Gandhi Nagar, Jammu, J&K, India.

Jamwal, G.S and Gupta, R.D. 2007 Watershed, management project- A case study of Bari-Badhori, J&K state (pp. 226-231). In Natural Resource Management for Sustainable Hill Agriculture (Eds. S. Arora, S.S. Kukal and V. Sharma) Soil Conservation Society of India, Jammu chapter, SKUAST Jammu, J&K, India.

Khajuria, G.L.2022. Need for greening wastelands. The State Times.27(293):6

Kumar, A Sharma, B.C and Sharma Vikas.2011. Sustainable Hill Agriculture—An Overview. Published by Agrobios (India), Agrohouse, Behind Nasarani Cinema Chopsani Road, Jodhpur Rajasthan.

Kumar, S. and Rattan, P. 2011. Vegetable Seed Production in Hills ln Sustainable Hill Agriculture An Overview (Eds. Anil Kumar, B.C. Sharma and Vikas Sharma) Published by Agrobios (India), Agro House Behind Nasrani Cinema, Chopasani Road Jodhpur, Rajasthan.

Kumar Parveen. 2022. Towards holistic development of Agriculture in J&K. The State Times. 27(252):6

Lal, Banarsi and Tandon Vikas, 2022. Boosting soil health by organic farming. Personnel Communication on the address of Professor, R.D Gupta, 258-C Sainik Colony, Jammu.

Raina, Jagdish Chander, 2021. Potential of organic farming in J&K. The Daily Excelsior. 57(328):6

Raina, S.K. and Chohan Abha. 2011. Natural Resource Conservation Technologies for Sustainable Crop produce. In Sustainable Hill Agriculture-An Overview (Eds. Anil Kumar, B.C. Sharma and Vikas Sharma), Published by Agrobio (India), Agro House Behind Nasrani Cinema Chopasani Road Jodhpur, Rajasthan.

Sabharwal, Gourav. 2022. Organic farming: Future opportunity and challenges in Union Territory of J&K. The State Times. 27(306):6.

Sharma, P. D. 2004. Managing natural resources in Indian Himalayas. Journal of Indian Society of Soil Science, 52(4)314-331.

Singh, M.P. 2007. Problems of micro and secondary nutrients in acid soils of India and their management. Bulletin of Indian Society & Soil Science. 25:27-58.

13

Agroforestry: An Organic Approach

Introduction

A number of experts have recognised three different stages in the march of man on the earth. These consist of: (i) When the civilization is dominated by the forests, (ii) When the civilization is overcoming the forest and (iii) When the civilization is dominating obstacles caused by forests. Now a days, Indian Himalayas including that of Jammu and Kashmir is passing through the third stage i.e., to the point of destroying itself in the process.

Deforestation's relentless march is stripping the Jammu and Kashmir Himalayas bare of its forest cover. From the majestic coniferous forests of Pir Panjal Himalayas to the deciduous and mixed canopy of subtropical forests the story is tragically the same i.e.. the indiscriminate destruction of forests that could turn the state into vast and inhospitable waste land. Hence to sustain and prosper the civilization, the forests must be saved from being further destroyed.

To reduce or to minimize the deforestation, the Social Forestry Project and Agroforestry System came to fore. The primary objective of Social Forestry is to increase supplies of fuel wood and fodder in rural area through plantation on community waste lands, degraded forest lands and on private lands through farm forestry (Patnaik, 1995-96). The latter i.e., Agroforestry is now much more widely used subject than the mere production of wood and food. In the changing scenario of the environmental degradation and ecological imbalance as a result of degradation of the forest resources of the country, agroforestry consists of inviting the prime attention of forests for upliftment of both environment and ecology vis-a-vis to enhance agricultural/horticultural and livestock production. In fact, agroforestry came as a new' name during the year 1978. Since then, the Indian Council of Agricultural Research (ICAR), New Delhi, has developed a vast network of agroforestry in all 36 centres of Agroclimatic Zones (Tejwani, 2008). However, Dr Tejwani started working with agroforestry during 1957 on fodder fuel plantations and trees on farm boundaries. Hence, agroforestry must be included in the agriculture.

It is because agriculture is the backbone of India's economy, which still contributes about 14.1 percent of the National income and furnishes with employment to around 6.4 per cent of the Indian working force (Debnath and Shrey, 2013). Further, it feeds 121 crores of human's population of the country directly or indirectly. The average per capita intake of cereals works out to be 427 grams per day during 2011-12, has remained satisfactory but there has been a fall in the per capita consumption of pulse is 35.8 grams per day in 2011-12. Thus, growing of pulses as component under agroforestry farming system must be emphasized. A part form this fish culture is also required to be included in the agroforestry system of farming, whenever it becomes possible (Chauhan, 2010).

An estimated 460 grams of cereal was available per person for each day across India in fiscal year 2022. Cereals included rice, wheat and maize among others. In financial year 2022, about 53 grams of pulses was available per capita daily in India.

Why Agroforestry

The forest resources have depleted considerably. This depletion has taken place during the last 136-137 years, and as such the rural community is facing serious problem in getting their basic necessities of life such as fuel wood, fodder, food, fruits, vegetables, grasses, flowers and timber etc. A villager has to traverse a long distance to collect a head load of fuel wood or fodder. However, this problem can be overcome by adopting agroforestry, which is a movement of the people, by the people and for the people. This can at least provide relief to the rural people by way of supplying the above stated requirement i.e., food, fodder, fuel wood timber etc. the movement of agroforestry has got international recognition spear-headed by FAO (Food and Agriculture Organisation) as reported by Rajdan (1995- 1996).

Purpose of Agroforestry/Objectives

1) To green the degraded lands/farms and villages to bring them in a natural environment and a balanced ecology. For this purpose, the vacant places are required to be covered with trees, bushes and grasses. These can include wasteland strips along streams, nallahs, religious places or panchayat houses and also those places where people usually sit to relax after the hard work done in the fields.

2) To obtain self- sufficiency in the primary needs of vegetables, fruits, dairy products, water for drinking and irrigating the crops fodder and fuel wood.

3) People's participation in the forest/community development is another objective.

4) To stabilize the rural economy, so that the fleeing people from the village are rendered to take interest in the above said works and discourage them from going in the cities.

5) To create employment potential of local people as labourers.

6) Educating the villagers about forestry ecology, environment both abiotic and biotic ones.

7) To obtain higher yields of food grain crops-rice, wheat maize, pulses etc, on the lines of foreign countries where production has increased by more than 50 percent through agroforestry. Thus, it will serve as food security.

8) To serve as a defence line against the destruction of natural forests by the rural people in light of the fact that these people shall have abundance of the produce by agroforestry.

Concept of Agroforestry

The concept of agroforestry formulates the technology of advantageously combining agriculture with forestry. As stated earlier that agroforestry is an innovation with far reaching significance and envisages the introduction of a green revolution in India's economy. Agroforestry, therefore, aims at diverse benefits to the farming community. These benefits may include

(i) Building up of farmers balanced economy and self-sufficiency. (ii) Providing fire wood, small timber and fodder and (iii) Earning extra source of income.

Agro forestry boosts the rural economy without impairing the quantity and quality of normal yields from the fields. The green revolution envisaged in agriculture cannot be achieved without the farmers self sufficiency and in fuel, fodder, manure and small constructional timber. Therefore, agroforestry is one among the means to attain and sustain green revolution target.

The concept of agroforestry in terms of its definition, aims & objective and uses, has been presented as under:

Definition of Agroforestry

"Agroforestry may be defined as a venture aiming at joint production of wood and food". However, now a day it has much more consideration of environments and ecology in the changing context of environmental degradation and ecological imbalance which have resulted from deforestation and consequently soil erosion, floods, denudation and devastation.

According to Sood (2019), agroforestry may be defined as an integrated land use system whereby trees, crops and or livestock are grown/reared on same unit of land.

Aims and Objectives

Farm forestry or Agro forestry is one of the components of the Social Forestry Project. Its main aim is to rehabilitate the degraded farms and villages through tree plantation on the farms, pasture lands, village waste lands and other land resources which are lying vacant. The broad objectives of agroforestry are:

i) To reduce the pressure of the villagers on protective and productive forests for making the local demands of fuel wood, fodder, timber, and other products, so that the existing forests can be fully spread for their protective and productive roles. In fact, agroforestry makes the villagers self-sufficient in fuel, wood, food and timber requirements. In other words, we can say that agroforestry is a defence line to protect the forests against their destruction by the rural communities.

ii) To utilise the available farm resources properly and more effectively.

iii) To maximise per unit production of food, fodder, fuel wood, livestock and other forest products. Besides, it should be raised with a view of optimising the productivity of biological and physical resources viz. Land, labour, livestock, soil moisture, solar radiation.

iv) To improve the natural environment and ecological balance of the degraded areas in terms of improving air quality and water.

v) To bring at least one third area under trees is to get more yields of crops and other agricultural produce on the lines of China where agroforestry has assisted in obtaining 50 per cent more increase in food grain production.

vi) To check the soil erosion and run off losses, conserve soil moisture and increase the soil fertility. For example, planting of *Saccharum munja* and *Saccharum spontaneum* on the bank of the rivers and streams to control the soil erosion caused by water and wind.

vii) To improve and stabilize the income of the farmer owing to increase in fuel wood and agricultural crops.

viii) To help in recycling of nutrients from lower soil horizons. The deep-rooted nature of most of the trees results in tapping of nutrients from deeper soil layers and returning them to plough layer through leaf drop and litter. Agroforestry leads to improvement of degraded lands and higher production from better lands by way of nutrient cycling, exploitation of deeper layers, nitrogen fixation, reducing demand on cow dung for fuel which in turn can be used as manure for crops.

ix) The estimated demand for food by 2025 will be 326 million tonnes (Prasad and Dhyani 2010) and food crops grown under agroforestry are likely to contribute 25 million tonnes per year.

Benefits of Agroforestry

i) Agroforestry system checks the rise in soil temperature especially during summer months. It, therefore, protects soil bioata (macro flora and macrofauna), which are essential for decomposition of organic matter and thereby, liberation of plant nutrients for crop production.

ii) Trees serve as shade for cattle and also improve the climatic conditions of the areas. For example growing of suitable plant species protect the farms from hot and cold winds and droughts in rural areas.

iii) It helps in utilization of run off season precipitation. Most of rainfed dry land areas are cropped for a single season but if a perennial tree component is there in the cropping system, it would make use of this efficiently.

iv) Tree component of agroforestry results in the improvement in the microclimate of the area.

v) In dryland agriculture agro-forestry helps in water conservation. The perennial component of the agroforestry system either on contour or in some other geometry helps in reduction of soil and water losses.

vi) Agroforestry imparts stability and results in reduction of two or more components of the system, even if one fails, the other would give a harvest. Thus a agroforestry decreases the risk of the crop failures.

vii) Agroforestry yields extended range of products viz. food, fodder, fuel, fiber, fruits, fertilizers. Presence of trees in agroforestry system can give extended management options also for the farmers i.e. trees can be harvested either for food or fuel or pole or small timbers as per requirements and market demands.

viii) Agroforestry makes up for loss of various tree species which has occurred due to deforestation, overgrazing and degradation of the forests.

Components of Agroforestry

Agroforestry possesses the following components:

i) Planting of forest tree species and shrubs alone or in combination with agriculture crops.

ii) Growing of fruit plants besides forest tree species/shrubs and agricultural crops.

iii) Growing of grasses (Table 6), fruit trees and forest trees besides agricultural crops.

iv) Raising of animals besides agriculture, horticulture, floriculture and forestry.

v) Growing of mulberry trees for rearing silkworm. This will boost the sericulture production.

vi) Rearing of fish in the ponds besides agriculture, horticulture and forestry. There must be significant interactions between the woody and non-woody components of the system either ecological and/or economical.

vii) This is a production system which tends to harmonies the production of various components and also maximized the total production from a given unit of land.

viii) The production and use are sustainable and makes use of modern technologies and traditional local experience besides being compatible with the social and cultural life of the local population.

ix) It is a long-term land management system and cycle of agroforestry system is always more than a year. In this way, agroforestry is a more complex form of land management both ecological and economically than other agricultural or forestry system,

x) Agroforestry creates forestry support for agriculture production including dairy, poultry, apiculture, horticulture vis-a-vis vegetables production.

Planting of Trees in India

In India, planting of trees in agroforestry by individual farmers seem to be the only solution to meet the challenge posed by the wide spread scarcities of fire wood, timber and fodder, and the growing in securities of our environments. Since the resources of the State Forest Department are inadequate to plant and maintain all the vacant private lands in the country under Rural Forestry Programme. Through agroforestry this responsibility is decentralised and the tree resources, thus, created are better protected and managed owning to individuals ownership(Table 4&5).

Table 4: Chemical composition of some forage trees.

Forage tree (Local name)	Botanical name	N	P	K	Ca	Mg
Khirk	*Celtis australis*	1.01	0.16	2.71	3.67	1.74
Dhaman	*Grewia optiva*	2.70	0.13	2.14	4.07	0.56
Kala sarin	*Albizia lebbeck*	3.14	0.12	2.42	4.08	1.07
Kau	*Olea cuspidata*	0.78	0.10	0.61	3.20	0.19
Khair	*Acacia catechu*	2.58	0.07	1.94	3.67	0.50
Bankhor	*Aesculus indica*	1.68	0.12	2.10	4.04	0.40
Ohi	*Albizia chinensis*	2.58	0.24	0.91	2.01	1.70
Su-babool	*Leucaena leucocephala*	2.74	0.14	2.64	2.48	0.70
Katrer	*Bauhinia variegata*	1.98	0.25	2.76	1.84	0.80
Phakari	*Ficus palmata*	2.40	0.25	1.92	3.41	0.59
Maggar bans	*Dendrocalamus hamiltonii*	2.34	0.20	2.01	0.84	0.70
Phaukarbor	*Zizyphyus nummularia*	2.31	0.13	3.10	3.11	0.64
Tut	*Morus alba*	2.34	0.29	2.90	3.71	1.42

Source: Gupta et al. 1985., Gupta and Arora, 2015.

Measures to Encourage the Agroforestry

To encourage the ecodevelopment or the rural areas through the following measures:

a) People's participation in forestry and local community development.

b) Interaction between stability in forestry and stability of rural community, enabling the village people not to migrate in towns/cities.

c) Encouraging stall feeding and creating fodder banks so that the people do not resort to free grazing in the forests.

d) Educating/creating awareness among the general masses regarding forestry/agroforestry environment and ecology, wildlife and hydrology.

e) Forestry for employment of the local people.

Advantages or Uses of Agroforestry/in Degraded Environment

There is a serious concern over degradation of environment. Serious adverse ecological manifestations viz. increase in carbon dioxide in atmosphere, change in climate, global warming, landslides, soil and runoff losses, flash floods, air, water and noise pollution, are the results of deforestation and overgrazing of the pastures.

i) An agroforestry system helps to increase the forest cover. It also makes available to people the required quantity of food, fodder,fuel wood, timber for which they depend upon forests. Thus, agroforestry system helps in reducing the pressure on forests and helps in conservation and development.

ii) Trees protect us from different kinds of pollutants. They protect us from dust, dirt and physical air pollutants. Record shows that one hectare of a close forest can filter about 50 tonnes of dust by the trees, protecting human beings from their adverse effects. Noise pollution is also checked considerably by the trees.

iii) The best control of landslides and soil erosion is through agroforestry i.e. planting of forest trees and grasses. Selection of tree species and grasses should however, be suited according to the locality. For example important species being planted in subtropical region of Jammu and Kashmir and Himachal Pradesh are: *Dalbergia sisso, A catechu, A modesta, Albizzia, lebbeck Cassia* spp., *Pinus roxburghii.*

iv) It has also been found that the deep rooted trees planted along contour and hedges along the cover crops help in soil stabilization and reduce landslide problems.

v) Agroforestry systems maintain soil fertility through recycling of nutrients and preventing soil erosion and loss of nutrients through leaching and runoff.

vi) Reduction in soil erosion and surface runoff help in reducing floods.

vii) In agroforestry system when leguminous trees are grown, they fix atmospheric nitrogen and return much more in leaf fall then they take from the soils.

viii) Leaves of the trees, could be used as a green manure, and help the farmers in increasing soil fertility. Agroforestry system are, therefore, helpful in sustaining land productivity at optimum level over a long period of time.

Improvement of Soil

Briefly, it is concluded that agroforestry systems help to improve soil in number of ways which include:

- Reduction of loss of soil as well as nutrients through reduction of runoff.
- Addition of organic matter through leaf fall, twig and bark fall.
- Nitrogen enrichment by fixation of nitrogen fixing trees, shrubs.
- Improvement of physical conditions of soils such as water holding capacity, permeability, cation exchange capacity and texture.

- Moderating the effect on extreme conditions of soil acidity, salinity, alkalinity and sodicity.
- Enhancing the microbial activities essential for the decomposition of organic matter.
- Lowering the effect on the water table in areas where the water table is high.

Trees, shrubs, herbs and climbers grown in agroforestry system would yield a considerable quantity of food and fodder which are used by the rural poor, tribals and others.

Fodder trees grown on large scale would act as fodder banks. Superior quality of grass species can be introduced for improving cattle health. (Table 3 and 4)

The trees and shrubs have a great advantage in dry areas over agricultural crops as they can withstand drought owing to their capability of tapping the lower layers of soil for water and nutrients. They increase employment opportunities for the rural poor and the landless people. Agricultural labour which is underemployed, finds alternative labour opportunities in agroforestry.

Agroforestry and Generation of Employment

Most of the agroforestry activities are labour intensive and there is considerable generation of employment opportunities. Plantations including nursery operations, generate employment of about 200 and 500 man days ha^{-1}. Subsequent care and maintenance also provide employment to the tune of 50 to 75 man days ha^{-1} $year^{-1}$.

Agroforestry systems are capable of meeting the demand of raw materials of several agricultural and forest based industries e.g. paper and pulp mills, sports goods, furniture, saw mills, are meeting their raw material requirements from forestry and agroforestry produce. For instance Populus species grown in Terai area of U. P. Parts of Punjab, Jammu and Kashmir, Himachal Pradesh and Haryana, are being used by a number of industries for furniture's, match splints, plywood, packing cases.

Agroforestry systems improve the productivity of plants and animals since they are based on sustainable land management and maximum utilization of natural resources to increase ecological and economic benefits.

Agroforestry is possible on uneven and undulating lands. Arid land, which is unsuitable for agriculture and, therefore, considered as non-productive, can be brought under tree plantation.

Evens small farmer can do boundary plantations. *Dalbergia sisso, Acacia modesta* can be planted on the bunds of the fields. The plantation of these species is useful for maintaining bee flora in apiculture. According to suitability of soils the poplars can be planted in the fields. In the space available in between the trees, herbal plants, oil seed crops can be raised.

Benefits of Agroforestry for the Rural Poor

In those areas where agricultural productivity is very low, but the demand for agricultural output is going up employment opportunities and potential are limited, there is a possibility of shifting the land use of agroforestry. While changing the land use system, it is important to see that production of food grains does not decrease correspondingly. If trees could be grown on intercrops and traditional food crops without affecting the crop yield substantially it would be acceptable to the rural poor.

Agroforestry systems are sustainable in drought prone areas. The grain yield in these areas is very low during the normal monsoon season and returns are very marginal during the drought affected year. Because of this reason, small and marginal farmers in these areas are poor and unable to increase their income if agroforestry is introduced in these areas the tree population will increase significantly although the grain yield of the field crops may go down to some extent. But during the years of drought when grain crops fail to generate any income, farmers can harvest the produce from trees for sustenance under the existing land use system, as most of the villagers are engaged in food grains production only. There exists little scope for getting higher wages through food production, especially in dry land farming areas. However, due to lack of alternative opportunities, most of the peasants spend their time in agriculture or remain in busy. Agroforestry can provide an additional employment in two ways.

a) Firstly, the farmers have to put an additional labour to maintain trees and to harvest the produce.

b) Secondly agroforestry can support new areas of employment.

Owing to uncertainty of rainfall and low returns, most of the small and marginal farmers particularly in drought prone areas are negligent and do not attend to the agriculture operations regularly. This leads to further reduction in yields of various crops. There is also degradation of land due to one or other reasons. By adopting agroforestry the farmers will tend to take good care of their crops and land, as they are sure of the profitability.

By introducing, planting of trees, timely intercultural operators will also result in soil and water conservation. Undulating topography, poor percolation of water due to tillage operations and poor vegetative cover, all these will contribute soil erosion. However, presence of trees/shrubs and grasses during agroforestry in agricultural fields can reduce soil erosion vis-a-vis help in reducing moisture evaporation.

Trees grown under agroforestry system also help in improving soil fertility by shading leaves (which on decomposition produce organic matter). They also increase soil moisture. Increased soil moisture helps in loosening the soil and partial shade assists in conserving humus in the soil.

Agroforestry in Jammu and Kashmir State

The state of Jammu and Kashmir located between 36° 58- 37° 17 N latitude and 73° 26 and 80 30' E longitude, has now been divided into two agroclimatic, geographical and administrative regions namely the Jammu and the valley of Kashmir. Each region of the J&K state is characterised by a different agroforestry system which stand described below:

As Major portion of Jammu and Kashmir state is hilly, where cool to extremely cool climatic conditions prevail during most part of the year. Moreover, most of the areas are far flung and as such are either inaccessible or linked with fair weather roads where regular supply of charcoal, coal and kerosene oil is not only difficult but is also very expensive. These areas also do not have any proper electricity facility. Thus, in such areas, the people have no option or alternative except to use fire wood for fuel purposes. They, therefore, obtain fallen material from the forests and where the fallen material is totally exhausted, they resort to felling and lopping of conifer trees. They also collect materials for their basic needs like fodder, timber, food, fruits, vegetables and flowers from the forests. This process cannot be stopped till some alternate arrangement for fire wood, and other needs is made available. Farm forestry or agroforestry is one of the alternatives to fill up this lacuna (Gupta, 2005; Gupta, 2009).

Two Important Multipurpose Agroforestry Trees for Kashmir

1) Honey Locust (***Gleditsia triacanthos*** L.) and Persian Lilac (*Melia azedarch*) are the most important multipurpose agroforestry trees grown in Kashmir. These are fast growing trees .The former tree with ornamental habit may reach up to 20 m in height. It has long trunk and attractively arranged branches densely covered with simple or compound flattened thorns. Leaves are pinnate with 20-30 leaflets or pinnate. Leaflets are acute to oblong, 2-3cm long, bright green, flowers are greenish which appear in June-July. Pods are flat stick shaped and twisted. The leaf fodder is nutritious and one lapping can be taken per season.

The nitrogen fixing capacity provides an edge to the trees. The tree bears pods or beans having sweetish succulent pulp with a sugar content of 27% which are relished by farm livestock and sometimes are eaten by human beings.

Among the advantages offered by the honey locust in agroforestry practice may be listed as rapid growth, improves the fertility of the soil owing to atmospheric nitrogen fixation, fodder value of leaves and pods, frost and drought hardy and thornless, 2) Melia azedarach (Persion Lilac): Afast growing tree is thought to be a native of Asia and possibly of North West India (Kashmir) but now it is naturalized throughout the tropics. It is found growing in the Himalayas up to 1800m above mean sea level. It is also found in Iran, Pakistan, Philippines, Cuba, China, Myanmar and Italy.

A Moderate sized tree with short bole and spreading crown. Bark is smooth, greenish brown when young turning grey and fissured with age. Leaves are delicate, fern like. Tree is leafless from December to March-April. Easily identified from panicles of Lilac-coloured flowers in March-May, occasionally also during Autumn. Fruits are soon formed in dark green clusters ripen during winter but remain on the tree till next April-May. The fruit is drupe, nearly globose yellow at first, wrinkled when ripe

Agroforestry Systems

As already stated, Agroforestry may be defined as a sustainable land management practice to increase the land utilization by integrating the production of crops and forest trees and/or animal rearing simultaneously or sequently on the same piece of land. Its main aim is to provide the rural poor with sufficient fuel food, fodder and timber on their farms to stop them raising the forests. This can be done only if agriculture and forestry go side by side by agroforestry system. In fact, this concept of growing multipurpose trees vis-a- vis crop cultivation on an agricultural land, farm bunds, waste and marginal lands has become a necessity for ensuring stability of the ecosystem and sustainability of production.

Agriculture - Silviculture, Silviculture-Pastoral, Agriculture-Horticulture-Silviculture are some of the agroforestry systems which are to be adopted by the farmers.

A) Jammu Region

Agriculture - Silviculture System: Studies undertaking at Dryland Agriculture Sub-Station, Rakh Dhainsar, Sher-e-Kashmir University of Agriculture Sciences and Technology (SKUAST) revealed that among the exotic trees

Leucaena - leucocephala and Leucaena diversifolia are promising for dryland areas of Jammu. The plantation of different indigenous species like *Grewia optiva, Acacia nilotica, A modesta, A catechu, Albizia lebbeck, Dalbergia sisso*, which provide fuelwood and small timber have been found useful in the Siwalik hills of Jammu. These species can be utilized suitably not only for agri-silviculture but for silvi-pastoral or horti-pastoral systems also for better productivity and conservation of natural resources. The important measures which need to be taken from soil and water conservation point of view in slopy lands include levelling, bunding, terracing of land, contour farming and mulching. Application of compost/manure and fertilizer should also be adopted. Preference must be given to add organic manures in view of organic farming. Crops like sesame and pulses (black gram, green gram) during kharif and gobi sarson (GSL-I) and toria are capable of doing well in agroforestry system of dryland areas of Kandi belt in Jammu.

In irrigated areas of R. S. Pura, *Prosopis juliflora* and *Acacia senegal* have shown their good growth. Hence, they must be included in this agroforestry system. *Sesbania aculeate*, in addition to serving green manuring, has also proved beneficial in providing fuel wood to the tune of 15 t ha^{-1} on dry weight basis about 150 days after its growth (Gupta et. al. 1992). Chemical composition of some of the forage trees in respect of nitrogen, phosphorus, potassium calcium and magnesium has been shown in Table 1.

Silviculture-Pastoral System: In this system of agroforestry, the trees are planted at wider spacing varying from 5x5m to 9x9m to provide leaf, fodder fuel wood and also allow the shrubs and grasses to grow. Tree species such as *Leucaena leucocephala* and *Grewia optiva* are proved useful. These are also effective in soil erosion. They act as shade tree, wind break and source of fence post and charcoal besides being a source of fodder and forest cover on dryland or rainfed agriculture.

Local shrubs such as *Adhatoda vasica, Carissa spinerum* and *Dodonaea viscosa* should be encouraged to grow in contour hedges to conserve soil and moisture. Local grasses will grow naturally but their grass slips can also be planted or sown through seeds. Introduction of grass species like *Stylosanthes hamata* can improve the soil fertility through atmospheric nitrogen fixation, while providng good quality fodder. This species is bushy in nature and thus protects the soil from erosion.

In field trials under dryland farming system, *Stylosanthes hamata* and *Stylosanthes scabra* have been found to fix 120 - 150kg of nitrogen /ha over a period of 3 years (Gupta et. al. 1992) average being 40-50 kg/ ha/ year.

Important local grass species are *Cynodon dactylon, Panicum antidotale* and *Chrysopogon fulvus*. It has been observed that *Cenchrus setigerus ciliaris* can effectively be grown in Kandi belt of Jammu. In loamy sand or sandy loam soils, planting of *Cenchrus* spp. gives good results. It is a point to mention that unpalatable bushes and shrubs like lantana camara, and *Parthinium* must be removed from the fields manually. These weeds are best controlled by tree planting. They disappear from shady areas.

Silviculture- Agriculture - Animal Husbandry System

In this system besides planting trees and grasses as stated in silviculture-pastoral system, some under planted crops can also be used wholly or in part to feed livestock and poultry. Sometimes domestic animals or poultry birds are allowed to feed directly on the harvested fields either between crops or on fallow land. In this agroforestry system, the excrements of the animals and poultry birds act as good source of manure for trees and crops.

Silvicultgure-Horticulture System: In this system besides fuel wood and fodder species, horticulture species are also planted. Horticulture plantations like lemon, orange, mango and ber can straight way be adopted in the Jammu Kandi belt in the light of their success in Dryland Horticulture Sub Research Station under SKUAST, Jammu and a number of other places of the area. Shade loving crops viz. turmeric (*Curcuma longa*) and taro or kachalu (Ecolocacia esculantao) were grown under *Grewia optiva*, *Bauhinia variegatae* and *Celtis australis*. Turmeric gave the highest rhizome yield (18.20q/ ha) under *Quercus leucotrichophora* followed by *Celtis australis* as reported by Gupta and Arora (2015).

Umran variety of ber has already been popularised in some of the Kandi areas like Raya. Suchani and Ranjri. A few Km ahead of the Eco Task Force catchment area in the Samba-Mansar Road, is the horticulture village of Dhora where a retired Army Captain Mr. Dhonda Singh started to grow citrus fruits (Kinnow) for the first time during 1984-85. Now almost each of about 30 families 'in the village are ex- servicemen, have found horticulture a far more remunerative ways than their traditional agriculture owing to the following reasons:

(i) There is an acute shortage of water in the kandi belt because of uncertain and erratic rainfall which cannot fulfil the requirement of agricultural crops. Horticultural crops, are, therefore, preferred due to their less water requirement. Some of the wild fruits, which can be grown in Kandi belt of Jammu especially in the surrounding areas of Mansar and

Surinsar lakes as recently prescribed by Singh(2019)) mainly consist of : 1) Bael (*Aegle marmelos*), 2) Amla (*Phyllanthus emblica*), 3) Ber (*Zizyphus mouritiana*), 4) Girna (*Carissa spinarum*), 5) Karonda (*Carissa carandus*) and 6) Lasura or Lasoda (*Cordia dichotoma*) etc.

(ii) Economic returns ha^{-1} of horticultural crops are higher than those of agricultural crops.

(iii) Horticultural crops ensure better soil protection.

The agroforestry started during 1962 by Mr. Sonaullah in wasteland area of Devgul village (Banihal, Jammu region in the Pir Panjal Himalayas), is worth mentioning. Today, this practice has spread over an area of 20Km in Banihal valley, involving 100 families of half dozen villages. Once totally wasteland, is now bringing about Rs.40,000-60,000' by sale of planted forest trees (Robina, Ulmus) and fruits (cherry, walnut, apple, almond, apricot, pear) and vegetables (cauliflower, cabbage, knolkhol, brinjal).

Silviculture - Agriculture - Sericulture - Animal Husbandry - Fishery

This is a very complex system of agroforestry together. It is characterised by a long path of exchange of material and flow of energy within an ecosystem. For example, if mulberry trees are planted, their leaves are used for silk worm rearing. The excrements of silk worms become good feed to the fish. The larvae of the fish are used as feed to the poultry. Silk is the end product.

In the mulberry stands, different crops can be under planted. Similarly, if forest trees are planted in a relatively wider space, a number of crops can be raised underneath for several years and can go to different uses, some for food, others for forage and still others for fruits and vegetable cultivation.

Jandra area of Jammu Shivalik's is found to be very congenial for cultivation of mulberry. First leaf harvest for silk worms rearing is taken for a year of planting.

Growing of French beans, and other vegetable like cauliflower, cabbage and knol-khol between the mulberry trees, have been found profitable. Fruits plants vis. Guava and Amla and grasses like *Stylosanthes* and *Setaria* can also be grown.

Silviculture- Agriculture -/Horticulture - Floriculture - Apiculture

It comprises production of flowers and honey besides growing of forest trees crops and fruit plants. Flowering plants often favour in the increase of parasites and predators of crop pests and thus as antiregulatory bio control system operates.

Under irrigated conditions, cultivation of marigold at 45 x45 cm apart has been found success fuel.

In between the row of marigold (*Tagets patula*), Onion (*Allium cepa*) was planted which has shown very good growth. Besides marigold, planting of rainy season annuals like balsam (*Impatiens balsamina*), chaulai (*Amaranthus caudatus*), cocks comb (*Celosia crisatata*) and flowering tree species like gulmohar (*Delonix regia*), bottle brush (*Callistemon*) have been found to grow very well. Apart from the above, some of the cash crops like zeera and saffron are also being cultivated. Cultivation of oats as a nutritive fodder is too being picked up. The improvement of the ecosystem can be judged by the following considerations:

i) Soil erosion of barren hill slopes has been checked by adopting agroforestry system.

ii) The birds and other wildlife species which had left, have reappeared on the rehabilitated land.

iii) There is lot of greenery in this part of Himalayan region.

The main fruit trees grown in irrigated areas of Jammu district are litchi, papaya, grapes, strawberry and citrus. The irrigated areas of Basohli and Bilawar are congenial for mango, litchi and guava cultivation. Temperate fruits like walnut, almond, apple, pear, apricot, peach and plum are grown in Bani and Lohai-Malhar area of Kathua, parts of Udhampur and Doda districts(Table 6).

Silviculture - Agriculture - Horticulture - Fishery

This system comprises raising of forest trees as well as agricultural/horticultural crops on the terraced lands, while ambient, ponds, tanks of lakes are good for fish culture.

The Kandi area of Jammu region popularly known as the land of ponds, has lot of prospects of this agroforestry system due to the presence of 336 ponds in this belt which can act as abode for fishery. For this purpose, the ponds which have dried must be revived.

Various species of fish abound in different parts of the aquatic ecosystem vertically. Fish species like silver carps are always found in the upper part and crucian carps close to the bottom with grass carps and black carps in between.

Lanceolatus spp, *Acacia modesta* and *A nilotica* have proved useful for this agroforestry system.

There are about 4500 reservoir fishermen in the state of Himachal Pradesh.

Similarly, there are lot of reservoir fishermen in Jammu and Kashmir state. The government provide them free nets so that they get some help in their fishing operation to promote fisheries in these states, the Govt, has been providing of incentives like subsidy of Rs. 80,000 ha-1 on creation of ponds in the agricultural fields and Rs. 15000 for renovation of ponds. The farmers must take an advantage of these schemes. Special emphasis is being laid on motivating farmers to take to commercial production of cold-water fish trout which is in great demand outside the state (Chauhan 2010).

It is worthwhile to mention that fisherman who will be entitled to these free fishing nets must fulfil certain condition like regular fishing for the last 3 years, must come forward to fulfil it and start fishing in ponds along with agriculture/ horticulture and forestry.

Kashmir Region

The farm forestry or agroforestry in Kashmir is an old practice or institution in the world. The history shows that farmers of Kashmir have been growing fruits plants, forest trees and legumes fodder since long in their agricultural fields.

The cultivation of temperate fruits in the valley is nearly three thousand years old, as has been mentioned by kalhana in the region of Nara (1000 BC).

Kalhana Says

"The forests were rich in wild and indigenous fruits and nuts trees of all sorts. These fruits and nuts trees provide food for wild animals in the forests. Some places are still known as Cheravan-Chuntwar-Tangi Dar. The fruits trees were grown on the sides of paths and sporadically in the agricultural fields and slopes. The regeneration was natural and it forms indigenous kind, and varieties available in abundance in and around forests. Planting a tree then, especially fruits or nuts tree, was considered a sacred act. When a boy head was shaved for the first time, the hairs thus cut used to be burried underground along with a few walnuts in the field or compound'*.

Long before the Moguls Kashmir rulers had laid out gardens or orchards in the valley. Grafting technique in horticulture was introduced from Central Asia. Some fruits like khotan-trel in Kashmir owe their origin to khotan and mongol-kushu like Ambri apple to central Asia. Similarly, a tall variety of Alfa-alfa owes its origin to yarkhand. In the late 19 century, experts from European countries viz. France and Italy were appointed during Dogra rulers. The Italians used to start grape cultivation for making wine. Mr. Paychaud, one of the French experts, remained in Kashmir valley and imported large number of fruit plants from France, United Kingdom along with a variety of flowering and ornamental plants. These plants were planted in the Maharaja's land around Dal Lake especially in the gardens of Raja Amar Singh at Naseem Bagh. Thereafter, fruit plant nurseries and orchards were established in whole of the Kashmir valley. The gardeners and staff of horticulture department were fully trained in the art of gardening and fruit cultivation of Shri Peychand, who headed the department of horticulture upto twenties.

Recently various steps have been taken to boost horticulture production. Apple is the most important temperate fruit and its production has increased by double in period of two decades. Walnut is the second largest fruit crop of this region and is important because it brings foreign exchange; almond is the third important fruit which is of much economic importance. Other fruits like plum, peach, cherry, pear and apricot are also grown. However, there is wide fluctuation in production of these fruits from year to year.

Agroforestry Systems

Following are the main agroforestry systems followed by the farmers.

Horticulture - Agriculture

The cropping intensity of Kashmir is by and large limited to single crop (rice or maize) only with horticulture as the main stay due to physiographic, climatic and other natural conditions. With the strides of Sher-e-Kashmir University of Agriculture Sciences and Technology, Kos'-1 variety of mustard is being grown. The production of wheat has almost become exhausted.

As already stated, that the Kashmir valley has a varying agroclimatic conditions which are most suitable for cultivation of apple, pear and other temperature fruits. The area is also very suitable for the production of off-season vegetables like cauliflower, cabbage, knolkhol, brinjals and many others during summer season for the plains.

Horticulture-Agriculture-Animal Husbandry

The raising of livestock in Kashmir besides production of horticultural and agricultural crops, is an important component of farming. This component can be divided into three categories viz., sedentary, semi migratory and migratory which are described hereunder

i) *Sedentary*: It is mostly practised in towns and cities. The livestock is mostly cattle which are reared in cattle shed. They are fed with green fodder during summer and rice straw during winter. Green fodder consists of scrub vegetation from forests and aquatic plants collected from lakes and other water bodies.

ii) *Semi migratory:* It is practised by the villagers dwelling in foot hills of the valley. One or two persons hired from village take the livestock of the village in pastures for grazing during summer. Meanwhile the villagers collect the grasses from forest for making hay. Willow tree leaves are also chopped and stacked on the tree. When the summer ends the livestock is then fed hay and tree leaves.

iii) *Migratory*: It is mostly adopted by the nomads locally known as Gujjars and Bakerwals. The livestock mostly consist of sheep goats and buffaloes, reach the pastures through nomads after melting snow of the hills during April - May and remain up to October. The nomads them start moving down and by November, they return at plains of Jammu, Punjab and Haryana. Animals graze in the aftermaths if paddy harvest in field and also by the paddy start purchased from the farmers. In February - March the again start moving towards hill pastures.

Agroforestry System in Cold Arid Region of Ladakh (Now Union Territory)

Within the cold arid zone of Ladakh, the agro climatic conditions vary greatly depending upon the altitude, kind of soil, climate season and cropping pattern. Accordingly, three agroforestry systems are quite prevalent in the area, which arc hereunder

Agriculture - Animal Husbandry

In this system of agroforestry, early maturing types of barley can be raised under irrigated conditions. This is suitable in the belt falling between 3450 to 4500m altitude which includes the sub-snow lines and high land meadow. Changthang, Khardun and Digger in Leh district and Zanskar and Dras in Kargil district fall under this belt.

Rearing of animals like goats, sheep and yak is equally important profession in this belt. Pastoral farming is patronized in Changthang sub division where special breeds of Pashmina goat and Changthangi sheep are reared by the nomads. Owing to very low temperature, no perennial forage crop is raised. Only local pea is cultivated for preservation as hay for winter feeding.

Agriculture-Horticulture/Olericulture - Animal Husbandry

This system is congenial for central belt of cold arid zone having elevation between 3000 to 3450m above mean sea level. This belt comprises mainly of Leh surroundings, Nobra valley and small part of Kargil.

Barely and wheat are the main cereals which are grown. Pea, mustard and lathyrus are also cultivated. Among fruits, apples and apricots are cultivated, though in limited scale. Alfa-alfa (lucerne) is the major forage crop raised and preserved in the form of hay for feeding to animals during November- May.

Animal rearing is a subsidiary profession. While adopting animal husbandry certain points like improved breeds, care of animals, preparation of balanced feed, control of parasites and common diseases, feeding of green fodder etc., are required to be followed on correct basis.

85mm rainfall in Leh and about 300mm in Kargil, hence no crop production is possible without irrigation. Among the cultivated fodders, alfa-alfa is highly adapted to the region and is most dependable fodder. Apple, raisin, grapes and apricot are the major fruit crops of the zone. Cauliflower, cabbage, knol khol and potato are the main vegetables grown. It is worthwhile to mention that Horticulture and Olericulture have undergone much change after 1947. One can notice now fruit trees grown even in small attached kitchen gardens.

Livestock rearing or pastoral farming is another major activity. Although total forest area of the region is practically zero, yet a large number of livestock such as sheep, goats and cattle are reared. During short summer alfa-alfa is extensively cultivated in irrigated area. The same is dried and preserved as hay for winter. The animals are also permitted to graze in open area (Gupta, 2009).

The pashmina goats are reared in Changthang belt on grazing in pastures which are situated along the river Indus. Besides fodder, lops and tops of willows and poplars grown in the common village lands on the outer skirts of Leh town, are offered to villagers as incentives to develop plantations. Three species of poplars viz *Populus candican, Populus euphratica* and *Populus nigra* have been planted in the village lands and along the river banks. Where irrigation water is assured. Populous candican has performed well in Leh range and *Populus eupharatics* in Nobra valley. The forest department is entrusted with the job of bringing more and more area under vegetation to bridge the gap between the demand and supply of fuels, to improve microclimate of the region to protect the exiting forests and check soil erosion. From 1988 and till 1992-93 the department has planted 18, 78, 295 plants of different species and 1 lakh during 1993- 94. These plants are now going on very well.

It is worthwhile to mention that agriculturally, the entire area is mono cropped except Khaltsi belt in Leh and Terchary in Nobra valley where double cropping is practised in abut 200ha. Cultivation is done at individual level in the rural and suburban areas fed by irrigation canals. Grim, wheat, lesser millets, pulses, are the main food products of these agricultural lands.

Raising of fast-growing forest tree species on village lands is very essential especially in cold arid zone especially in Ladakh Union Territory and Lahaul Spiti in Himachal Pradesh. For this purpose, planting of *Robinia pseudoacacia* has proved very useful. It not only provides fodder, fuel, wood but also fixes atmospheric nitrogen in soils.

Homestead Agroforestry

Some of the rattans (Apotential source of Cane) like Daemonorops Jenkinsinus, Calamus khasianus and Calamus guruba are traditionally part of homestead garden in north east hill region. They can be encouraged to grow for a more scientific homestead agro forestry.

Although the impact of declining rattan resources had adversely affected the Jhumias (Jhume cultivators) who are the traditional gatherer of rattan from wild sources for cash income, yet they can grow some rattan species by following a new agro forestry system.

New Agroforestry System

In this Agroforestry system, rattan can be incorporated in a short rotation commercial forestry with Kadam (*Anthocephalus cadamba*) avenue plantation and many small land holder production systems to diversify farm enterprises to raise their income. Lesson can be learnt from Malayasia to develop rubber + rattan, cultivation.

The above said approaches require merit attention since rattans cannot be grown under monoculture and even rare species can be rehabilitated through agroforestry. Thus, productive conservation can be the best approach of biodiversity management of rattan species of North-east India, and hence agroforestry will prove itself to be a powerful intervention.

References

Bandral, B. S. 1997. Trial of major range grasses of Jammu and Kashmir. Forest News Letter 3(Nov. 1996 - January 97).

Bhushan, C. Shukla, A., Sahoo, B and Vishwanath, 2006. Strategy for maximization of forage production in Uttranchal Hills Indian Farmers Digest. 39 (8): 9-11 and 14.

Debanath, P and Shrey, R. 2013. Food Security in India. Agrobios News Letter. 12(4):80-81.

Gupta, R. D., Gupta, J. P. and Singh, H. 1992. Problems of agriculture in the Kashmir Himalayas. In conserving Indian Environment (Edited by S. K. Chadha). Pointer Publishers, Jaipur, Rajasthan (India).

Gupta, R.D., Gupta, S.K. and Mahajan A/ 2022. Sustainable system of Farming for Modern Agriculture NIPA Genx Electronic Resources & Solutions P.Ltd, New Delhi

Gupta, R.D. 2003. Food production potential through agroforestry of Jammu Kandi Belt. The Kashmir Times. 60(288):7

Gupta, R. D. 2005. Agroforestry and Enrichment of Environment. In Environmental Degradation of Jammu and Kashmir Himalays and Their Control (Edited by R. D. Gupta), Associated Publishing Company, Post Bo\ No. 2679, Karol Bagh New Delhi - 110005 India.

Gupta, R. D. 2005. Environmental Degradation of Jammu and Kashmir Himalayas and their control. Associated Publishing Company, Karol Bagh New Delhi, India.

Gupta, R. D. 2009. Glimpses of North West, Himalayas. Radhakrishna Anand and Co. Pacca, Danga Jammu Tawi, 180001, J&K India.

Gupta, R.D. 2012. Agro forestry of Ladakh. The daily Excelsior. 48(230):6

Gupta, R. D. and Abrol, V. 2006. Agroforestry - A sound land use system for Himalayan rainfed ecosystem Farmers Forum. 6(3): 19-22.

Gupta, R. D. and Arora, S. 2015. Agroforestry as alternative land use system for sustaining rural livelihoods in Himalayan ecosystem. Advances in soil and water resource management for food and livelihood security in changing climate.

Gupta, R.D. Marwah, B.C. And Bali, S.V. (1985). Soils of Jammu and Kashmir and their management. In soils of India and their management, FA, New Delhi.

Lal Banarsi and Sharma Pawan 2021. Empowering farmers of J&K through dairy techniques. The State Times 26(150):6

Patnaik. P. 1995-96. People's participation in social forestry programme in Jammu and Kashmir. Dharti. 4:1-3

Prasad, R. And Dhyani, S.K. 2010. Review of Agroforestry contribution and problems in arid ecosystem for livelihood support in India. Journal Soil and Water conservation. 9(4):277-287.

Razdan, A. K. 1995-96. Agroforestry - A look Dharti Publication- 4:12-13

Sharma, G. K. 1993-94. A brief note on concerted efforts made by social forestry project to meet the requirement of locals living in the vicinity of forage production area (Devak Bella). Dharti Publication (3):57-60.

Singh, R. 2019. Wild fruits of Mansar Surinsar. The Daily Excelsior Magazine. 55(54):4 [Sunday, February 24]

Sood, Kamal Kishore. 2019. Agro forestry for boosting farmers income. The State Times.. 24(216):6

Tejwani, K.G. 2008. India 60: Contribution of Agroforestry to economy, livelihood and environment (Part i). Wastelands News: 23 (4):32-33.

Appendices

Appendix A

List of Indian Magazines Related to this book having Title "Organic Farming for Sustainable Agriculture"

Agriculture Today	Monthly	Centre for Agriculture and Rural Development 502, Rohit House, Telstoy Road, New Delhi
Agrobios Newsletter	Monthly	Agrobios (India) Behind Nasrani Cinema Chopasani Road, Jodhpur
Command Area Development Subjar	Quarterly	Command Area Development Talab Tilo, Jammu, J&K
Cooperative Bulletin	Monthly	General Secretary J&K Cooperative Union, Vir Marg, Jammu, J&K
Down to Earth	Monthly	Centre for Science and Environment, New Delhi
Dharti	Annual	Soil Conservation Training Sector, Miran Shahib, Jammu, Jammu & Kashmir
Environmental and People	Monthly	Society for Environment of Education Hyderabad-500073
Employment News	Weekly	General Manager and Chief Editor, East Block (V), Level-7 R.K. Puram, New Delhi-110086
Farmers Forum	Monthly	Bharat Krishak Samaj, New Delhi
Farmers Parliament	Monthly	Executive Committee of the Farmers, Parliamentary Forum, New Delhi
Farm Progressve	Monthly	Associated Publisher, Madras Ltd. 856, Annasaloi
Gram Vikas Jyoti	Quarterly	Council for Development of Rural Areas 2, Vigyan Lok, Vikas Marg, Extension, Delhi
Indian Farmers Digest	Monthly	Direction Communication Centre, G.B. Pant University of Agriculture and Technology, Pant Nagar, Uttarakhand
Indian Farming	Monthly	Indian council of Agricultural Research Krishi Anusandhan Bhavan, New Delhi
Intensive Agriculture	Monthly	Krish Vistar Bhavan, Directorate of Extension Ministry of Agriculture, Pusa, New Delhi
Progressive Farming	Monthly	Punjab Agriculture University Ludhiana, Punjab

Appendix B

List of Indian News Papers, related this book

The Tribune	Daily	Printed and published (J&K Edition) Tribune Trust at Danik Jagran priting Press Bari Brahmna Samba, J&K
The Daily excelsior	Daily	Printed & published by Mr Kamal Rometra and owned by daily Excelsior Excutive Editor Mr Neeraj Rometra Pritnted at Daily Excelsior Excelsior Printer Pvt.Ltd. Janipur, jammu-180007, J&K
The State Times	Daily	Printed & Published at State Time Publications, Plot No 40, 111-112, Phase-II, Gangyal, Jammu-180007
The Kashmir Times	Daily	Printed and Published by Parmodh Jamwal for and on behalf of Kashmir Times Publication, Printed at KT PressPvt. Ltd. at Digana and Published from Residency Road, Jammu, Jammu and Kashmir

Appendix C

Methods of organic Farming: Organic farming involves fostering natural methods or processes with the objective to enhance the soil health. A variety of methods can be utilized which mainly comprise of using of organic manures including organic fertilizers and green manuring.

Organic manures are classified into several ways. However, precisely speaking organic manure is the one that is the result of some natural product such as farm yard manure, compost, vermicompost etc., containing all the three essential plant nutrients (NPK) required to be supplied to the soil in varying quantities along with secondary (Ca, Mg, S) and micronutrients (Fe, Mn, Cu, Zn, Mo B and Cl) in smaller quantities.

Appendix D

Nutrient composition of organic manures and organic fertilizers

Farmyard manure (FYM)				
S.No	Organic Manure	Nutrients Percentage		
1.	Farmyard manure (FYM)	N	P2O5	K2O
2.	Compost	1.00	0.60	3.00
Vermicompost Organic Fertilizer				
1.	Guano	16.30	1.94	5.11
2.	Molasses	0.24	0.07	12.2
3.	Dried blood	12.70	2.00	1.81
4.	Oil Cakes	5.40	1.70	1.40

Appendix E

Some of the biofertilizers are also called of living bacteria which have been in use since long for improving crop yield. These are applied to seeds roots of pants or to the soils as an inoculants

Appendix F

The inoculation of *Rhizobium* to legumes has now become an established practice in several counties like U.S.S.R.; U.S.A., France, Belgium, Sweden etc. The use of such culture has also become in India during recent years. Other bacterial fertilizers prepared from *Azotobacter* bacterial species and *Bacillus megatherium* var *phosphaticum* are the other biofertilizers used these days as *Azotobacterin* and *Phosphobacterin* inoculants/cultures.

Appendix G

The other methods which are followed in organic farming consist of as cover cropping pattern as well as using of mulching practice

Appendix H

In organic farming, there is always use of biopesticides instead of chemical pesticides. It is resorted to lessen the pesticidal pollution which has taken place not only in the air but also in the water and soil.

Table 5: Main Fodder Tree Species for Planting in Various Zones of Jammu and Kashmir State.

Agroclimatic Zone	Tree Species and Bushes
Subtropical Submontane of Foot Hilla of Shivalik's (300-1000m)	*Grewia optiva, Albizia lebbeck, Bauhinia spp. (B. variegata, B. purpurea), Acacia spp. (A. nilotica, A. catechu, A. modesta). Butea monosperma, Celtis australis, Morus alba, Carissa spinarum* Corida dichotoma, Ficus religiosa, F. glomerata. Bombax cieba.*
Mid Hills of Shivalik's of Middle Mountains of Inner Himalayan Region (1000-1500m)	*Grewia optiva, Celtis australis, Robinia monospherma, Ficus palmata, Albizia procera, Albizzia chinensis, Populus spp. Quercus leucotrichophora, Morus alba, M. austrails, M. nugra.*
Higher Himalayas or Great Himalayan Region (1500- 2000m)	*Robinia pseudocacia, Celtis australis,Morus alba, Quercus leucotrichophora, Salix spp.,Pupulus spp.*
Cold Arid Zone (2000- 2500m)	*Betula utilis, Salix spp. Prunus spp. Populus ciliata, P. alba. P. candicans, P. euphratica*

Source: Gupta, R. D. 2005; Gupta, R. D. 2009., Gupta, R. D. and Abrol, V 2006, Gupta and Arora, 2015

Table 6: Fruit , Medicinal and Ornamental Trees

Common Name	**Scientific Name**
Ber	*Ziziphus mauritiana*
Mango	*Mangifera indica*
Pomegranate	*Punica granatum*
Guava	*Psidium guajava*
Jamun	*Syzygium cumini*
Phalsa	*Grewuia asiatica*
Lasoora	*Cordia dichotoma*
Mulberry	*Morus alba*
Papaya	*Carica papaya*
Citrus fruits	*Various species*
Medicinal Trees	
Harar	*Terminalia chebula*
Bahera	*Terminalia belerica*
AmlaNeem	*Emblica officinalis or Phyllanthus emblica* *Azadirachta indica*
Corniferous & Broadleaved Species	
Chir or Chil	*Pinus roxburghii*
Kail	*Pinus wallichiana*
Deodar	*Cedrus deodara*
Banj oak	*Quercus leucotrichophora*
Moru oak	*Quercus floribunda*
Ornamental Trees	
Araucaria	*Araucaria cookie*
Bottle brush	*Callistemon lanceolatus*
Weeping willow	*Salix babylanica*
Amaltas	*Cassia fistula*
Rubber plant	*Ficus elastica var* Black Prince Bauhinia variegata *B blakeane*
Gulmohar	*Delonix regia*
Silver oak	*Grevilla robusta*
Ahok	*Polyathia longifolia*
Arjun	*Terminalia arjuna*

Index